Lecture Notes in Computer Science 15225

The series Lecture Notes in Computer Science (LNCS), including its subseries Lecture Notes in Artificial Intelligence (LNAI) and Lecture Notes in Bioinformatics (LNBI), has established itself as a medium for the publication of new developments in computer science and information technology research, teaching, and education.

LNCS enjoys close cooperation with the computer science R & D community, the series counts many renowned academics among its volume editors and paper authors, and collaborates with prestigious societies. Its mission is to serve this international community by providing an invaluable service, mainly focused on the publication of conference and workshop proceedings and postproceedings. LNCS commenced publication in 1973.

Xiaobo Zhou · Chen Yu · Song Guo ·
Jianping Wang · Xianhua Song · Zeguang Lu
Editors

Green, Pervasive, and Cloud Computing

19th International Conference, GPC 2024
Macao, China, September 27–30, 2024
Proceedings

Springer

Editors
Xiaobo Zhou
University of Macau
Macau, China

Chen Yu
Huazhong University of Science
and Technology
Wuhan, China

Song Guo
The Hong Kong University of Science
and Technology
Hongkong, China

Jianping Wang
City University of Hong Kong
Kowloon Tong, China

Xianhua Song
Harbin University of Science and Technology
Harbin, China

Zeguang Lu
National Academy of Guo Ding Institute
of Data Science
Beijing, China

ISSN 0302-9743 ISSN 1611-3349 (electronic)
Lecture Notes in Computer Science
ISBN 978-981-95-1345-1 ISBN 978-981-95-1346-8 (eBook)
https://doi.org/10.1007/978-981-95-1346-8

This Springer imprint is published by the registered company Springer Nature Singapore Pte Ltd.
The registered company address is: 152 Beach Road, #21-01/04 Gateway East, Singapore 189721, Singapore

Preface

As the program chairs of the 19th International Conference on Green, Pervasive, and Cloud Computing (GPC 2024), it was our great pleasure to welcome you to attend the conference, which was held in Macao, China, during 27th–30th September 2024, hosted by the University of Macau, Huazhong University of Science and Technology, Harbin Institute of Technology and HeiLongJiang Computer Federation and National Academy of Guo Ding Institute of Data Sciences. The goal of this conference was to provide a forum for researchers and practitioners to discuss and exchange novel ideas, results, experiences and work in progress on Green, Pervasive, and Cloud Computing.

This conference attracted 33 paper submissions. After the hard work of the Program Committee, 13 papers were accepted to be presented at the conference, with an acceptance rate of 39.4%. There were at least 3 double-blind reviewers for each article, and each reviewer reviewed no more than 5 manuscripts. The major topic of GPC 2024 was Green, Pervasive, and Cloud Computing. The accepted papers cover a wide range of areas related to Cyber-Physical-Social Systems, Edge Intelligence, Industrial Digitization and Applications, Mobile Sensing and Computing and Wireless and Ubiquitous Networking.

We would like to thank all the program committee members of the conference for their hard work in completing the review tasks. Their collective efforts made it possible to attain quality reviews for all the submissions within a few weeks. Their diverse expertise in each research area helped us to create an exciting program for the conference. Their comments and advice helped the authors to improve the quality of their papers and gain deeper insights.

Great thanks should also go to the authors and participants for their tremendous support in making the conference a success. Besides the technical program, this year GPC offered different experiences to the participants. We hope you enjoyed the conference.

June 2025

Chen Yu
Song Guo
Jianping Wang

Organization

General Chair

Xiaobo Zhou	University of Macau, China

Program Chairs

Chen Yu	Huazhong University of Science and Technology, China
Song Guo	The Hong Kong University of Science and Technology, China
Jianping Wang	City University of Hong Kong, China

Steering Committee Chair

Hai Jin	Huazhong University of Science and Technology, China

Steering Committee

Cho-Li Wang	University of Hong Kong, China
Zhiwen Yu	Harbin Engineering University, China
Nabil Abdennadher	University of Applied Sciences, Western Switzerland, Switzerland
Christophe Cerin	Sorbonne Paris Nord University, France
Sajal K. Das	Missouri University of Science and Technology, USA
Jean-Luc Gaudiot	University of California-Irvine, USA
Kuan-Ching Li	Providence University, Taiwan
Chao-Tung Yang	Tunghai University, Taiwan
Laurence T. Yang	St. Francis Xavier University, Canada/Hainan University, China

Honorary Chair

Rajkumar Buyya	University of Melbourne, Australia

Publication Chair

Xianhua Song	Harbin University of Science and Technology, China

Local Organizing Chair

Xiaowei Wu	University of Macau, China

Finance Chair

Pengyang Wang	University of Macau, China

Program Committee Members

Alex Vieira	Federal University of Juiz de Fora, Brazil
Andrea Abate	University of Salerno, Italy
Anupama Mishra	Swami Rama Himalayan University, India
Arcangelo Castiglione	University of Salerno, Italy
Baoliu Ye	Nanjing University, China
Bihuan Chen	Fudan University, China
Boyuan Ma	University of Science and Technology Beijing, China
Carson K.Leung	University of Manitoba, Canada
Chen Liu	City University of Hong Kong, China
Chen Ma	City University of Hong Kong, China
Chunhua Su	Aizu University, Japan
Cong Wang	City University of Hong Kong, China
Dan Li	Tsinghua University, China
Dingqi Yang	University of Macau, China
Federica Sarro	University College London, UK
Gangyong Jia	Hangzhou Dianzi University, China
Gautam Srivastava	Brandon University, Canada

Gongming Zhao	University of Science and Technology of China, China
Guojie Luo	Peking University, China
Guoqiang Li	Shanghai Jiao Tong University, China
Han Yu	Shanghai Jiao Tong University, China
Hao Zhou	University of Science and Technology of China, China
Hausan Wong	City University of Hong Kong, China
Hongkai Wen	University of Warwick, UK
Hui Zhang	University of Science and Technology of China, China
Jian Li	Shanghai Jiao Tong University, China
Jianguo Yao	Shanghai Jiao Tong University, China
Jianli zhao	Shandong University of Science and Technology, China
Jingyu Hua	Nanjing University, China
Jinlei Jiang	Tsinghua University, China
John C. S. Lui	Chinese University of Hong Kong, China
Jorge Barbosa	Universidade do Porto, Portugal
Jun Huang	City University of Hong Kong, China
Junqiao Qiu	City University of Hong Kong, China
Ka-Chun Wong	City University of Hong Kong, China
Kaigui Bian	Peking University, China
Kanchana Thilakarathna	University of Sydney, Australia
Kang Chen	Tsinghua University, China
Kuo-Chan Huang	National Taichung University of Education, Taiwan
Lei Bu	Nanjing University, China
Lei Xie	Nanjing University, China
Li Li	University of Macau, China
Mahmut T. Kandemir	Penn State, USA
Mario Marino	Leeds Beckett University, UK
Minghui Zhou	Peking University,China
Minming Li	City University of Hong Kong, China
Nan Guan	City University of Hong Kong, China
Ning Chen	City University of Hong Kong, China
Peiquan Jin	University of Science and Technology of China, China
Yaokai Feng	Kyushu University, Japan
Qingchuan Zhao	City University of Hong Kong, China
Qiong Luo	Hong Kong University of Science and Technology, China

Quanfeng Deng	Huazhong University of Science and Technology, China
Shih-Wei Li	National Taiwan University, Taiwan
Shiqi Wang	City University of Hong Kong, China
Shyam Burkule	Meta, Inc, USA
Tianzhang Xing	Northwest University, China
W. K. Chan	City University of Hong Kong, China
Wasim Ahmad	University of Glasgow, UK
Wei Hu	Nanjing University, China
Wei Tong	Nanjing University, China
Wei Xu	Tsinghua University, China
Weidong Yang	Fudan University, Chia
Weitao Xu	City University of Hong Kong, China
Wenfei Wu	Peking University, China
Wenzhong Li	Nanjing University, China
Xiaofeng Chen	Xidian University, China
Xiaoheng Deng	Central South University, China
Xiaotong Wu	Nanjing Normal University, China
Xiapu Luo	Hong Kong Polytechnic University, China
Xin Niu	Huazhong University of Science and Technology, China
Xin Wang	Fudan University, China
Yang Gu	Chinese Academy of Sciences, China
Yangfan Zhou	Fudan University, China
Yanmin Zhu	Shanghai Jiao Tong University, China
Yipeng Qin	Cardiff University, UK
Yuan Wu	University of Macau, China
Yuan Zhang	Fudan University, China
Yuezhi Zhou	Tsinghua University, China
Yuntao Wang	Tsinghua University, China
Zengwei Zheng	Hangzhou City University, China
Zhaoguo Wang	Shanghai Jiao Tong University, China
Zhenjiang Li	City University of Hong Kong, China
Zhi Yang	Peking University, China
Zhicong Lu	City University of Hong Kong, China
Zhijie Wang	Chongqing University, China
Zili Shao	Chinese University of Hong Kong, China

Contents

Cyber-Physical-Social Systems

Image-Text Sentiment Analysis Based on Cross-Modal Interactive Attention

Mairidan Wushouer[1], Xu Guo[2](✉), Gulanbaier Tuerhong[1], Hao Huang[2], Liwei Tian[1], Suping Liu[1], and Longqing Zhang[1]

[1] Guangdong University of Science and Technology, Dongguan, China
mairidan@gdust.edu.cn, gulan@gdsut.edu.cn
[2] Xinjiang University, Urumqi, China
guo_xu_up@163.com, huanghao@xju.edu.cn

Abstract. This paper introduces a novel approach to improving performance through the integration of cross-modal interaction and attention mechanisms, specifically channel and spatial attention for sentiment analysis tasks. By focusing on the synergistic relationship between textual and visual information, our proposed Cross-Modal Interaction Attention (CMIA) model leverages the strengths of both data types to achieve superior classification accuracy. In this paper we detail the methodology of channel and spatial attention mechanisms, demonstrating how they prioritize relevant features and spatial locations within the data. Through a series of experiments, including an ablation study on the MVSA dataset, we show that our model not only outperforms existing approaches either traditional or SOTA, but also highlights the importance of cross-modal interactions for a deeper understanding of sentiment. Our findings suggest significant potential for the application of attention mechanisms in various domains, pushing the boundaries of current neural network capabilities in sentiment analysis.

Keywords: Multimodal Sentiment Analysis · Multimodal Fusion · Attention Mechanism

1 Introduction

The social media platforms have become a predominant venue for the exchange of information, witnessing an unprecedented surge in the volume of multimodal content. This proliferation of textual and visual data is a rich reservoir of emotional cues, encapsulating diverse sentiments such as joy, anger, and sadness, as well as nuanced attitudes and behavioral intentions. Understanding these multifaceted emotional states through multimodal sentiment analysis is pivotal for a comprehensive appraisal of user sentiments and intentions. The dynamic interplay between different modalities—text and images, in particular—offers a fertile ground for enhancing the accuracy of sentiment analysis.

X. Zhou et al. (Eds.): GPC 2024, LNCS 15225, pp. 3–16, 2026.
https://doi.org/10.1007/978-981-95-1346-8_1

Multimodal data, characterized by the inherent complementarity between text and visual information, poses a unique challenge for sentiment analysis. This complementarity can significantly enhance the interpretation of sentiments conveyed through multimodal content. For instance, an isolated examination of textual or visual modalities may not yield an accurate sentiment inference, as illustrated in Fig. 1. The figure exemplifies scenarios where the integration of text and images is instrumental in discerning the true sentiment, underscoring the limitations of uni-modal analysis. This complementary nature is not merely additive but involves intricate correspondences between textual descriptions and visual elements, necessitating a nuanced understanding of cross-modal interactions to accurately interpret multimodal sentiments. In recent years, a prevalent trend among researchers has been the utilization of attention mechanisms to strengthen the intricate connections between varying types of data [1, 2].

To address these challenges, this paper introduces a novel approach centered on Cross-Modal Interaction Attention (CMIA). This methodology leverages dual-attention mechanisms—channel and spatial attention within visual data—to augment the representational capability of ResNet50 [3]. Furthermore, it employs cross-modal interactive attention to map the associations between image regions and textual semantics, thereby enriching the model's comprehension of multimodal content.

The efficacy of CMIA is evaluated using two multimodal sentiment datasets: MVSA-Single and MVSA-Multiple. Our experimental results demonstrate that CMIA outperforms several baseline models across these datasets, substantiated by improvements in accuracy and F1 score metrics. Through a series of ablation studies, the superior performance of CMIA in multimodal fusion is validated, underscoring its potential in advancing the field of multimodal sentiment analysis. The contributions of this work are twofold:

1) We introduce a Cross-Modal Interaction Attention (CMIA) framework that integrates dual-attention mechanisms with cross-modal interactive attention, enhancing the synergy between textual and visual modalities.
2) The proposed CMIA model is validated on the MVSA-Single and MVSA-Multiple datasets, showcasing its superiority in multimodal sentiment analysis as evidenced by improved accuracy and F1 scores compared to baseline models.

This study not only advances the methodology of multimodal sentiment analysis but also sheds light on the importance of leveraging cross-modal interactions, setting the stage for future research in the domain.

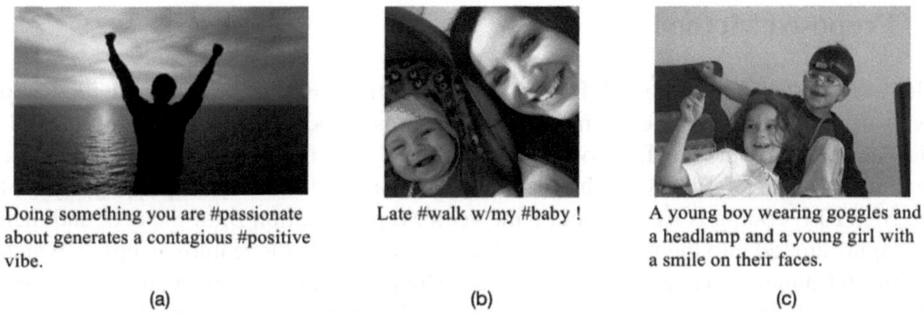

Fig. 1. A sample Image-text pair from the MVSA dataset. (a) The words passionate and positive in the text convey positive emotions, but the images have no obvious emotional tendencies, (b) The smile in the image expresses a positive emotion, but the text has no specific words that express emotional tendencies. (c) Words in the text such as young boy and goggles correspond to certain regions in the image. Human emotions are primarily evoked by key words and key areas in images.

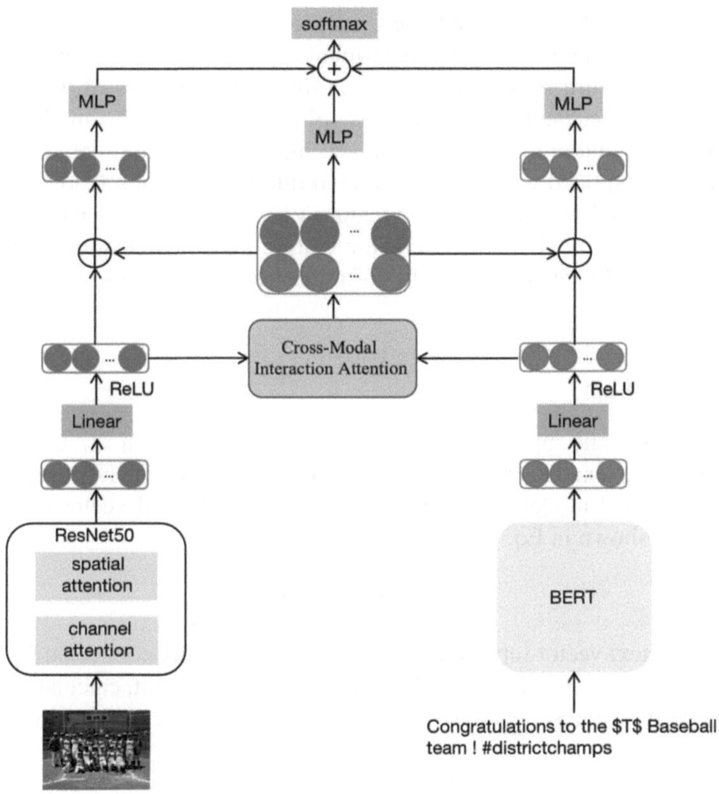

Fig. 2. CMIA Overall Architecture.

2 Proposed Method

The overall structure of the CMIA model is shown in Fig. 2, which is divided into two parts: the feature extraction module and the cross-modal interaction attention module.The inputs of the CMIA model include text X and image I, and the outputs are sentiment scores ψ. Firstly, in the feature extraction module, the last pooling layer of the BERT [4] coding is used to obtain the vector representation B_i of the text, and then the inputs are fed into the fully-connected layer to complete the feature mapping, which outputs 64-dimensional text features T_i; for the images I of the comment, this paper uses ResNet50-CBAM to extract the image features. ResNet50-CBAM introduces channel attention and spatial attention into ResNet50 and uses the output of the last convolutional layer of its network as the image representation F_i, and then input F_i fed into the fully-connected layer to obtain the 64-dimensional image V_i. Second, in the cross-modal interactive attention module, drawing on the idea of Transformer [5] structure, the multi-head attention mechanism is used to realize text-image mutual guidance to highlight key emotional features, and the text-image interactive fusion feature H_{t-v} and the image-text interactive fusion feature H_{v-t} are obtained, and then the interactive features are inputted into the feed-forward network layer and residual block to obtain the final cross-modal interactive fusion feature H. In order to retain the useful features of the original text and the original image, the text feature T_i is spliced with the cross-modal interaction fusion feature H to obtain the text-cross-modal interaction fusion feature TC; and the image feature V_i is spliced with the cross-modal interaction fusion feature H to obtain the image-cross-modal interaction fusion feature VC. Then, the cross-modal interaction feature H, text-cross-modal interaction fusion feature TC and image-cross-modal interaction fusion feature VC are input into the softmax classifier to obtain the final sentiment score ψ.

2.1 Modal Encoding

Text Encoding. For the input text data represented as $X = \{X_0, X_1, \ldots, X_i, \ldots X_n\}$, where i denotes the i text data in the text-image dataset, the text data is processed through the BERT pre-trained model to output a 768-dimensional word vector representation denoted as B_i, as shown in Eq. 1.

$$B_i = BERT(X_i) \tag{1}$$

The obtained text vector representation B_i is fed into a fully connected layer for feature mapping, resulting in a 64-dimensional text feature as output, calculated according to Eq. 2, where W_h^T represents the weight coefficients and b_h represents the bias term.

$$T_i = W_h^T B_i + b_h \tag{2}$$

Image Encoding. Channel attention focuses on dynamically adjusting the importance of different channels in the input feature map, so that the model can learn useful feature information more effectively. The structure of channel attention is shown in Fig. 3.

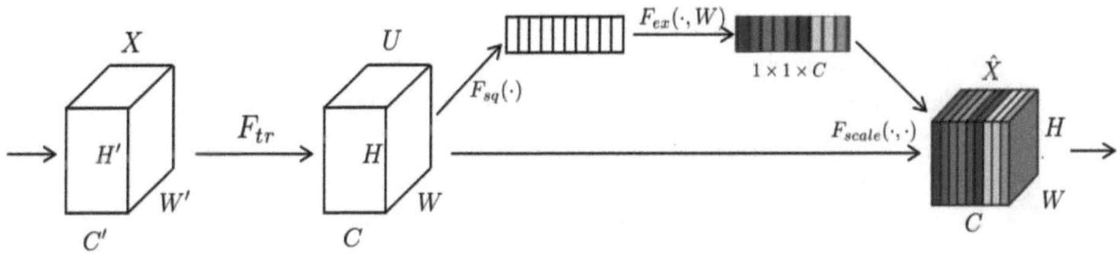

Fig. 3. Structure of channel attention

Spatial attention refers to an attention mechanism that improves the model performance and generalization ability by focusing on the information of different spatial locations in the image when processing the image. The structure of spatial attention is shown in Fig. 4.

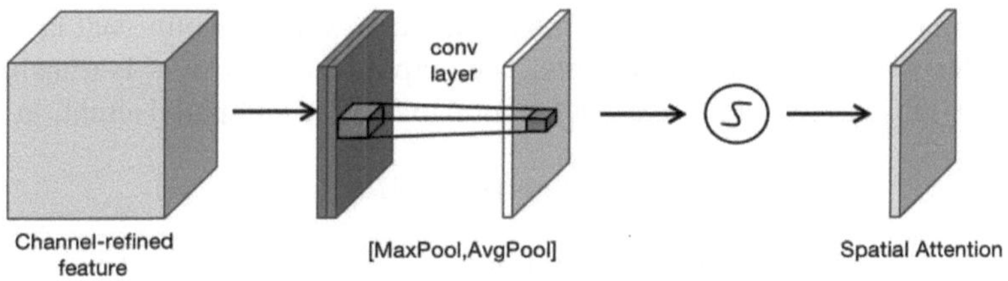

Fig. 4. Structure of spatial attention

For image feature extraction, this paper employs ResNet50 in conjunction with the Convolutional Block Attention Module [6] (CBAM), herein referred to as ResNet50-CBAM. The architecture of ResNet50-CBAM is delineated in Fig. 5.

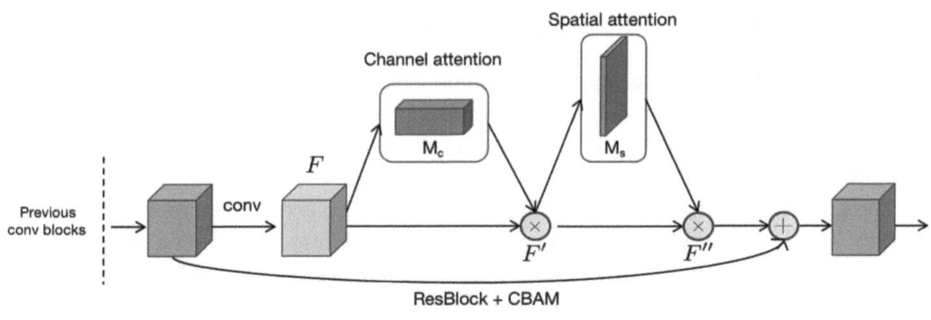

Fig. 5. Structure diagram of ResNet50-CBAM

For the images $I = \{I_0, I_1, \ldots, I_i, \ldots I_n\}$, where I_i denotes the image in the dataset associated with the i th textual data, firstly, the channel attention and spatial attention modules are applied after a specific layer of ResNet50 and the attention-weighted features are multiplied with the original features by element-level multiplication to obtain the output of the channel attention and spatial attention processing.

The channel attention module accepts the output I_i from the 4th residual block of ResNet50, following convolution and pooling operations, as its input. During channel attention computation, the input features' average and maximum values are computed using adaptive average pooling and adaptive maximum pooling, respectively. These values are subsequently channelled through two parallel fully connected layers, which utilize 1×1 convolutions for the purpose of learning channel-wise attentional features. The outputs from these layers are aggregated, and the final channel attention weights are determined using the sigmoid activation function, as delineated in Eq. 3.

$$CA_i = \sigma(FC_{avg}(AvgPool(I_i)) + FC_{\max}(MaxPool(I_i))) \tag{3}$$

where σ is the Sigmoid activation function and FC_{avg} and FC_{max} denote the mean and maximum fully connected layers, respectively.

Following channel attention processing, the spatial attention mechanism is engaged as in Formula 4. It commences by crafting a two-channel feature map derived from the mean and maximum values across the input features. A convolutional layer is then applied to learn spatial attention weights. The output from this layer is transformed to the range (0, 1) via a sigmoid activation function to finalize the spatial attention weights.

$$SA_i = \sigma(Conv([AvgPool(CA_i), MaxPool(CA_i)])) \tag{4}$$

where *Conv* represents convolution operation.

The spatial attention weights, once derived, are then processed through a fully connected layer and an activation function to attain the final image feature representation V_i. This representation is dimensionally constrained to 64, as delineated in Formula 5.

$$V_i = ReLU(FC(SA_i)) \tag{5}$$

2.2 Cross-Modal Interactive Attention Module

The cross-modal interactive attention module fuses text and image features using a Encoder-like structure in Transformer. It harnesses a multi-head attention mechanism to fortify connections and complementary capacities across different modalities. Figure 6 displays the Cross-modal Interactive Attention Module.

The CMIA module produces dual interaction features: text-guided image interaction features and image-guided text interaction features. These are derived through text-to-image and image-to-text association models, respectively. The multi-head attention mechanism empowers the model to focus on pertinent elements during image and text processing, which in turn enriches the semantic links between them.

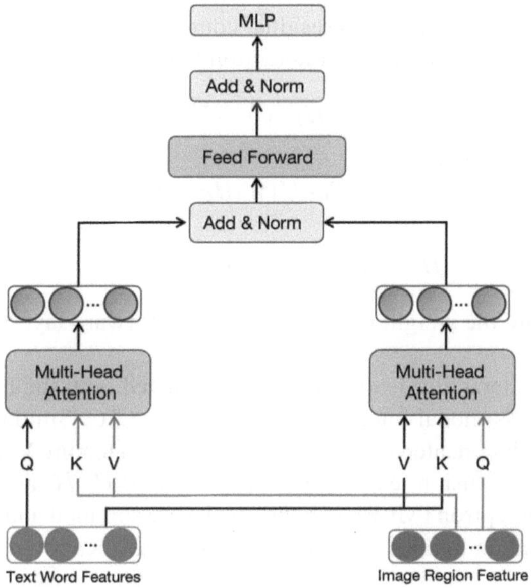

Fig. 6. Structure of Cross-modal Interactive Attention Module.

The interaction features between text-guided image features utilize the cross-modal interaction of text word features guiding image region features. This model allows text features to guide image features, enabling them to interact in a cross-modal scenario. This interaction tightens the correlation between image understanding and text comprehension, thereby aiding in enhancing the model's performance in cross-modal tasks. Taking text-guided image interaction features as an example, the calculation formula for the i-th attention head is shown in Eq. 6, where H_T represents the query of text features, w_q^i,w_k^i and w_v^i represent the projection parameters in the i-th attention head, h denotes the number of attention heads, and d_h represents the feature dimension of input features for parallel attention heads.

$$Head_i(H_T, H_v) = SoftMax\left(\frac{\left[w_q^i H_T\right]^T \left[w_k^i H_v\right]}{\sqrt{dh}}\right)\left[w_v^i H_v\right]^T \tag{6}$$

After the parallel computation of each attention head, their outputs are concatenated and then subjected to a linear transformation to derive the multi-head attention output. This process is outlined in Eq. 7, where W_m denotes the parameters of the linear transformation within the multi-head attention framework, and h denotes the count of attention heads.

$$MultiHead(H_T, H_v) = W_m(Head_1(H_T, H_V), ..., Head_h(H_T, H_V)) \tag{7}$$

The output from each attention head is utilized to generate the text-guided interactive image feature $\mathrm{H_{T-V}}$, and likewise, the image-guided interactive text feature $\mathrm{H_{V-S}}$. These resultant features are then amalgamated and processed through a feedforward neural network layer. This is succeeded by a residual connection and layer normalization step to form the comprehensive image-text interaction feature H.

$$H_c = H_{T-V} \oplus H_{V-T} \tag{8}$$

$$H_{FFN} = W_2(ReLU(W_1 H_C + b_1)) + b_2 \tag{9}$$

$$H = LayerNorm(H_c + H_{FFN}) \tag{10}$$

where W_1 and W_2 are the weight matrices of the feedforward layers and b_1 and b_2 are the bias vectors.

The text-image interaction feature H is concatenated with the text word feature T to obtain the text-cross-modal interaction fusion feature TC. Similarly, the text-image interaction feature H is concatenated with the image region feature V to obtain the image-cross-modal interaction fusion feature VC. These features H, TC and VC are then input into a multi-layer perceptron (MLP), and the results are summed and processed through a SoftMax normalization to obtain the final emotion classification result ψ.

$$TC = H \oplus T \tag{11}$$

$$VC = H \oplus V \tag{12}$$

$$\psi = SoftMax(MLP(H) + MLP(TC) + MLP(VC)) \tag{13}$$

3 Experiment and Performance Analysis

This section evaluates the effectiveness of the feature extraction as well as the cross-modal interaction attention modules through comprehensive experiments. The implantation was written in Python 3.7 on Ubuntu 20.04.6 and the Pytorch 1.13.0 [7] was used as deep learning framework. The training process was accelerated using GeForce RTX 3090 GPUs.

3.1 Datasets

We evaluated the proposed model on the public multimodal sentiment analysis dataset MVSA [8], which is composed of two subsets: MVSA-Single and MVSA-Multiple. These subsets encompass image-text pairs across three sentiment categories: positive, neutral, and negative. The MVSA-Single dataset contains 5,129 image-text pairs, each annotated with an emotional label by a single annotator. Conversely, the MVSA-Multiple dataset includes 19,600 image-text pairs, each annotated by three independent annotators for emotional content. Image-text pairs with inconsistent annotations have been excluded to ensure comparability. Details of the MVSA-Single and MVSA-Multiple datasets are outlined in Table 1.

Table 1. The detail of MVSA multimodal dataset.

Dataset	Positive	Neutral	Negative	totals
MVSA-Single	2683	470	1358	4511
MVSA-Multiple	11318	4408	1298	17024

3.2 Evaluation Metrics

We use accuracy and F1 score as evaluation metrics, which are defined as follows:

$$\text{Precision} = \frac{TP}{TP + FP} \tag{14}$$

$$\text{Recall} = \frac{TP}{TP + FN} \tag{15}$$

$$F1 = \frac{2 \cdot \text{Precision} \cdot \text{Recall}}{\text{Precision} + \text{Recall}} \tag{16}$$

where TP represents the number of samples correctly labelled as positive examples, FP represents the number of samples incorrectly labelled as positive examples but were actually negative examples.Recall is used to measure the model's ability to correctly predict all actual positive samples.

3.3 Experimental Settings

We fine-tuned the hyperparameters to achieve optimal performance on the validation set. The training dataset and validation dataset are divided in the ratio of 8:2.The pre-trained BERT base model's parameters are kept static, with a text feature dimension set at 768. Image dimensions are standardized to 224 × 224 before being input into the ResNet50-CBAM network for visual feature extraction, resulting in a feature dimension of 2048. Text data processed through BERT and image data processed through ResNet50 both undergo a linear transformation to a dimensionality of 64. Adam's algorithm is utilized for gradient-based optimization. To prevent overfitting, dropout rates are set at 0.2 for bert_dropout, 0.2 for resnet_dropout, and 0.5 for fuse_dropout. Additional hyperparameters are detailed in Table 2.

Table 2. Values of hyperparameters.

Hyperparameters	Settings
optimizer	Adam
learning rate	5e-5
batch size	16

(continued)

Table 2. (*continued*)

Hyperparameters	Settings
attention_nhead	8
attention_dropout	0.4
fuse_dropout	0.5
middle_hidden_size	64
loss_function	CrossEntropyLoss
epoch	20
weight_decay	0.01

3.4 Experimental Results and Comparison Study

To assess our method's efficacy, comparative experiments were conducted on sentiment classification tasks against both traditional and recent works. The models compared are:

MultiSentiNet [9]: Utilizes visual semantic information extracted from objects and scenes within images, guiding an LSTM model with visual features to extract keywords via attention, which informs classification and labeling.

Co-MN [10]: Introduces a co-memory network to iteratively model interactions between images and text to enhance multimodal sentiment classification.

Table 3. Comparison of classification performance of baseline models on the MVSA dataset

Method	MVSA-Single		MVSA-Multiple	
	Accuracy	F1	Accuracy	F1
MultiSentiNet	69.84	69.63	68.86	68.11
CoMN	70.51	70.01	69.92	69.83
D-Senti	70.07	69.14	72.03	69.2
DMAF	70.29	70.03	68.86	66.76
MVAN	72.98	72.98	72.36	72.3
MMBT	73.17	72.43	72.21	70.63
MMBT-F	73.39	73.14	72.39	69.86
MGNNS	73.77	72.7	72.49	69.34
TSAFNet	74.28	73.19	72.91	70.95
CGAFT	74.77	74.56	72.98	71.96
ITIN	75.19	74.97	73.52	73.49
CLMLF	75.33	73.46	72	69.83
Our(CMIA)	76.63	74.09	75.03	71.48

D-Senti [11]: D-Senti connects visual and textual features to predict sentiment labels.

DMAF [12]: DMAF is a deep multimodal attention fusion approach to image-text sentiment analysis.

MVAN [13]: A multimodal sentiment analysis framework based on multiview attentional networks, leveraging memory network modules to iteratively garner semantic image-text features.

MMBT [14]: Developed by Facebook AI Research, this model utilizes multimodal bi-directional transformers for image-text classification, integrating text and image inputs.

MMBT-f [15]: Adapts MMBT for sentiment analysis on imbalanced graphic-text datasets.

MGNNS [16]: Proposes a multi-channel graph neural network leveraging global dataset features for learning image-text representations.

TSAFNet [17]: An attention-based two-stage neural network designed for the sentiment classification of text and visual content.

CLMLF [18] (Contrastive Multimodal Learning Framework): Employs label-based and data-based contrastive learning tasks to assist the model in learning common features related to emotions in image-text pairs.

CGAFT [19] (Cross-Modal Graph Attention Fusion Transformer): Offers an approach for multimodal sentiment analysis, with adaptive fine-tuning of the visual Transformer and cross-attention fusion of patch-level image features with text-level text features.

ITIN [20]: ITIN proposes a new image-text interaction network to study the relationship between emotional image regions and text words for image-text sentiment analysis.

The detailed comparative results presented in Table 3 allow us to draw the following observations:

(1) MultiSentiNet achieved an accuracy of 69.84%, acknowledging the impact of visual elements on text. CoMN, which learns from the iterative interactions between visual and textual elements, marginally surpassed MultiSentiNet. MVAN, which leverages both object and scene features in images for cross-modal learning, outperformed the previous baselines. The ITIN model, with its cross-modal alignment and gating modules, facilitated deeper text-image interactions, thus securing higher accuracy and F1 scores.
(2) The CMIA model introduced in this paper surpasses the earlier MultiSentiNet by 6.79% in accuracy and 4.46% in F1 score on the MVSA-Single dataset. It also excels beyond the current leading CLMLF by margins of 1.3% in accuracy and 0.63% in F1 score. Notably, the CMIA model's F1 score lags behind CGAFT and ITIN, potentially due to data imbalances. On the MVSA-Multiple dataset, CMIA improves upon MultiSentiNet's results by 6.17% in accuracy and 3.37% in F1 score and also edges ahead of the top-performing ITIN by 1.51% in accuracy, though it falls slightly short in F1 score comparison. The research considers the potential for model bias towards certain categories.

Overall, the findings affirm the benefits of the CMIA approach in graphical sentiment analysis. The enhancement in performance is credited to the robust feature extraction method and the cross-modal interaction mechanism. The feature extraction module efficiently isolates critical information from both texts and images, while the cross-modal interaction module delves into the semantic correlations between them, extracting valuable and complementary information that leads to more precise sentiment predictions.

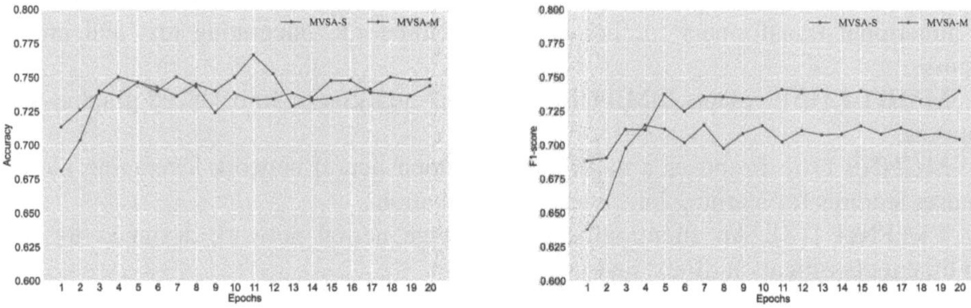

Fig. 7. Variation of model accuracy and F1 values during training

Figure 7 illustrates the progression of accuracy and F1 scores on the validation set with increasing training epochs on the MVSA dataset. On the graphic, accuracy metrics are displayed on the left, with F1 scores on the right. The MVSA-Single dataset is represented by the red curve, while the MVSA-Multiple dataset is indicated by the blue curve. The data indicates a consistent improvement in both accuracy and F1 scores for the MVSA-Single and MVSA-Multiple datasets as the number of training epochs rises. Peak accuracy for the MVSA-Single dataset is reached at epoch 11, whereas for the MVSA-Multiple dataset, the apex occurs at epoch 7, demonstrating rapid model convergence. The trends affirm that the model persistently learns and refines its parameters throughout the training, leading to enhanced sentiment analysis performance. As training advances, the model increasingly aligns with the distinct features of the dataset, ultimately achieving stable and reliable performance. This convergence reflects the efficacy of the training regimen and the model's capability to attain commendable results on the dataset.

3.5 Ablation Study

To assess the significance of each component within the model, an ablation study was conducted. Starting with the full model architecture, the cross-modal interactive attention module was first removed. Subsequently, text and image features were concatenated using a basic fusion method and input into a softmax classifier to perform sentiment analysis. Furthermore, the study examined the individual contributions of text and image features to the model's performance by conducting separate sentiment analysis for each modality. The results of these ablation experiments, which aim to delve into the model's inner workings and the influence of each module, are presented in Tables 4.

By substituting the sophisticated cross-modal interaction attention module with a simpler concatenation approach, the study can discern the specific contribution of the cross-modal interaction attention to the model. Additionally, isolating text and image sentiment analysis provides insights into the relevance of each modality to the overall sentiment analysis task, guiding further optimization of the model's design and performance.

Table 4. Results of the ablation study of the MVSA dataset

Method	MVSA-Single		MVSA-Multiple	
	Accuracy	F1	Accuracy	F1
Without image	70.50	69.87	70.61	70.01
Without text	70.25	69.79	69.40	68.27
Without Cross-modal interaction module	71.75	70.73	71.58	70.54
CMIA	76.63	74.09	75.03	71.48

The ablation study revealed that the CMIA full model delivered the most superior classification outcomes, validating the integral role of each component within the CMIA framework in enhancing overall performance. Results indicated that even when employing only text or image features, high accuracy was attainable, demonstrating the pre-trained model's prowess in extracting pivotal feature information from both modalities. When text and image features were combined, there was a marginal improvement in performance, underscoring the added value of multimodal data over unimodal data. Nonetheless, this combination alone was insufficient for capturing the intricate correlations between text and images. The cross-modal interaction module, introduced in this paper, is instrumental in facilitating deeper and more effective interaction between text and image modalities, as reflected in the performance metrics.

4 Conclusion

In this paper we have presented a comprehensive exploration of the role of attention mechanisms in enhancing the performance of sentiment analysis tasks, with a particular focus on the integration of cross-modal interactions. Our proposed CMIA model effectively combines channel and spatial attention to prioritize and process the most relevant information from both textual and visual inputs. The experimental results, particularly the ablation study on the MVSA dataset, underscore the importance of cross-modal interaction in achieving high classification accuracy and model efficiency. By leveraging the complementary strengths of text and image features, our model not only advances the state-of-the-art in sentiment analysis but also opens new avenues for research into attention mechanisms and their applications across various fields. The success of the CMIA model points to potential of attention-based and cross-modal approaches in enhancing the interpretability, performance, and generalization capabilities of neural networks, setting a new benchmark for future work in the domain.

References

1. Sun, L., et al.: Efficient multimodal transformer with dual-level feature restoration for robust multimodal sentiment analysis. IEEE Trans. Affect. Comput. (2023)
2. Wu, J., et al.: A optimized bert for multimodal sentiment analysis. ACM Trans. Multimed. Comput. Commun. Appl. **19**(2s), 1–12 (2023)
3. He, K., et al.: Deep residual learning for image recognition. In: Proceedings of the IEEE Conference on Computer Vision and Pattern Recognition, pp. 770–778 (2016)
4. Devlin, J., et al.: BERT: pre-training of deep bidirectional transformers for language understanding. arXiv preprint: arXiv:1810.04805 (2018)
5. Vaswani, A., et al.: Attention is all you need. In: Advances in Neural Information Processing Systems, vol. 30 (2017)
6. Woo, S., et al.: CBAM: convolutional block attention module. In: Proceedings of the European Conference on Computer Vision (ECCV), pp. 3–19 (2018)
7. Paszke, A., et al.: Automatic differentiation in pytorch (2017)
8. Niu, T., et al.: Sentiment analysis on multi-view social data. In: MultiMedia Modeling: 22nd International Conference, MMM 2016, Miami, FL, USA, January 4–6, 2016, Proceedings, Part II 22. Springer International Publishing, pp. 15–27 (2016)
9. Xu, N., Mao, W.: MultiSentiNet: a deep semantic network for multimodal sentiment analysis. In: Proceedings of the 2017 ACM on Conference on Information and Knowledge Management, pp. 2399–2402 (2017)
10. Xu, N., Mao, W., Chen, G.: A co-memory network for multimodal sentiment analysis. In: The 41st International ACM SIGIR Conference on Research Development in Information Retrieval, pp. 929–932 (2018)
11. Hu, A., Flaxman, S.: Multimodal sentiment analysis to explore the structure of emotions. In: Proceedings of the 24th ACM SIGKDD International Conference on Knowledge Discovery & Data Mining, pp. 350–358 (2018)
12. Huang, F., et al.: Image–text sentiment analysis via deep multimodal attentive fusion. Knowl.-Based Syst. **167**, 26–37 (2019)
13. Yang, X., et al.: Image-text multimodal emotion classification via multi-view attentional network. IEEE Trans. Multimedia **23**, 4014–4026 (2020)
14. Kiela, D., et al.: Supervised multimodal bitransformers for classifying images and text. arXiv preprint: arXiv:1909.02950 (2019)
15. Basu, P., et al.: Multimodal sentiment analysis of# metoo tweets using focal loss (grand challenge). In: 2020 IEEE Sixth International Conference on Multimedia Big Data (BigMM), pp. 461–465. IEEE (2020)
16. Yang, X., et al.: Multimodal sentiment detection based on multi-channel graph neural networks. In: Proceedings of the 59th Annual Meeting of the Association for Computational Linguistics and the 11th International Joint Conference on Natural Language Processing (Volume 1: Long Papers), pp. 328–339 (2021)
17. Hu, X., Yamamura, M.: Two-stage attention-based fusion neural network for image-text sentiment classification. In: Proceedings of the 2022 4th International Conference on Image, Video and Signal Processing, pp. 1–7 (2022)
18. Li, Z., et al.: CLMLF: A contrastive learning and multi-layer fusion method for multimodal sentiment detection. arXiv preprint: arXiv:2204.05515 (2022)
19. Xiao, X., et al.: Image–text sentiment analysis via context guided adaptive fine-tuning transformer. Neural Process. Lett. **55**(3), 2103–2125 (2023)
20. Zhu, T., et al.: Multimodal sentiment analysis with image-text interaction network. IEEE Trans. Multimedia (2022)

Research on Robot Path and Joint Angle Optimisation Based on the A* and Simulated Annealing Algorithms

Yong Tian(✉), Guona Chen, and Yongjie Chen

China University of Petroleum-Beijing at Karamay, Karamay, Xinjiang, China
2022015945@st.cupk.edu.cn

Abstract. With the continuous advancement of automation technology, six-degree-of-freedom robotic arms are becoming more and more widely used in manufacturing, medical care, logistics, and services. Thanks to their flexibility and efficiency, these robots are gradually replacing traditional manual operations, improving production efficiency and reducing labour costs. However, how to achieve efficient path planning and joint angle optimisation for robots in complex and ever-changing environments remains a major challenge. To address this problem, this paper proposes a comprehensive application solution that combines the A algorithm with the simulated annealing algorithm. The A algorithm uses a heuristic search method that can quickly find the optimal path in a large state space, making it particularly suitable for multi-objective path planning in dynamic environments. The algorithm improves the speed and accuracy of path planning by comprehensively considering actual and heuristic costs. It is highly adaptable and can be adjusted in real time to meet different task requirements. The simulated annealing algorithm, as a global optimization tool, can effectively avoid local optima and help achieve an optimal configuration of joint angles. The algorithm performs well in high-dimensional optimization problems, improves the movement accuracy and efficiency of the robotic arm, and reduces energy consumption. Experimental results show that the path planning and joint angle optimisation strategy combining A* and simulated annealing algorithms can effectively cope with the challenges in complex environments and improve the overall performance of the robotic arm. The research results of this paper provide theoretical support and practical guidance for the application of intelligent robots in complex environments, and have important academic and practical value. Future research can explore the combined application of other optimisation algorithms to further enhance the adaptability of robotic arms in a wider range of scenarios, so as to promote the sustainable development and innovation of robotics.

Keywords: Robotic Arm · Joint Angle Path · Optimal Design · A* Algorithm · Simulated Annealing Algorithm

X. Zhou et al. (Eds.): GPC 2024, LNCS 15225, pp. 17–27, 2026.
https://doi.org/10.1007/978-981-95-1346-8_2

1 Introduction

With the rapid development of automation technology, robots are being used more and more widely in various industries, especially in manufacturing, medical care, logistics and services. Six-degree-of-freedom robotic arms are ideal for complex operations due to their flexibility, precision and efficiency. These robotic arms can perform a variety of tasks, such as assembly, handling, welding, medical surgery and fine manipulation, greatly improving work efficiency and operational precision. However, how to achieve efficient path planning and joint angle optimisation in complex and dynamic environments remains an important issue that needs to be addressed urgently.

In a dynamic and ever-changing environment, the robot arm must not only adapt to different task requirements, but also effectively respond to the distribution and changes of obstacles. Therefore, real-time path planning is particularly important [1]. Traditional path planning methods such as Dijkstra's algorithm and A* algorithm have been widely used for this type of problem, but the efficiency and adaptability of these methods when dealing with dynamic environments are often limited. Although Dijkstra's algorithm can theoretically find the optimal solution, its computational complexity is relatively high, especially when the number of nodes is large, and it is not real-time. The introduction of the A* algorithm has provided a new way of thinking to solve this problem. Using a heuristic search method, it can quickly find the optimal path in a large state space, making it particularly suitable for path planning in dynamic environments.

In the process of path planning, in addition to finding the optimal path, the optimisation of joint angles is also crucial. The joint angles of the robotic arm directly affect the accuracy and efficiency of its movement. Traditional methods for joint angle optimisation are based on analytical solutions or inverse kinematics, but these methods require high model accuracy and are not flexible enough [2]. In recent years, global optimisation algorithms such as genetic algorithms, particle swarm optimisation algorithms and simulated annealing algorithms have gradually attracted attention. These methods can effectively avoid local optima and help achieve the optimal configuration of joint angles, and are particularly suitable for complex multi-peak optimisation problems [3].

This study aims to explore the combined application of A* and simulated annealing algorithms to achieve path planning and joint angle optimisation for a robotic arm in a complex environment [4]. We will analyse the applicability of these algorithms in multi-objective path planning and joint angle adjustment, and evaluate their impact on improving the efficiency of robotic arm movement and grasping accuracy. By comparing and combining various algorithms, we hope to support the application of intelligent robots in dynamic and uncertain environments in theory and practice.

Specifically, this paper will first review the existing work related to this research and discuss the main algorithms and techniques for current path planning and joint angle optimisation. Next, we will describe in detail the basic principles, implementation steps and combination methods of the A* algorithm and simulated annealing algorithm [5–10]. Finally, the effectiveness and practicality of the proposed method will be verified through practical experiments, with a view to laying the foundation for future robot technology development and promoting its application prospects in more fields.

Through this comprehensive research, we hope to provide new ideas for real-time path planning and optimisation of robotic arms, not only improving their operational efficiency, but also enhancing their ability to cope with various challenges in complex environments. This will help to achieve a more intelligent robotic system and promote the further development of Industry 4.0 and intelligent manufacturing.

2 Related Work

In the research on robot path planning and joint angle optimization, scholars have extensively improved existing algorithms. Early path planning methods, such as the Dijkstra and A* algorithms, laid the foundation for this field. While Dijkstra performs well in undirected graphs, its efficiency in dynamic environments with complex obstacles is limited [11]. The A* algorithm, with its heuristic search advantages, has become a key focus, leading to various improvements, such as enhanced heuristic functions and dynamic update mechanisms for better performance in dynamic environments. For joint angle optimization, traditional methods based on analytical solutions or inverse kinematics lack flexibility in real-time tasks, prompting researchers to adopt global optimization algorithms like genetic algorithms, particle swarm optimization, and simulated annealing [12]. These methods excel in complex, multi-peak problems by leveraging population search, randomization, and mechanisms like temperature control to escape local optima and find global solutions.

In recent years, research combining multiple optimisation algorithms has gradually increased. Scholars have attempted to combine local search with global search in order to improve the overall performance of the algorithm. For example, some research combines the A* algorithm with the genetic algorithm to form a new hybrid algorithm, which aims to solve the path planning problem in a dynamic environment. In addition, the combination of simulated annealing with other local search strategies has also shown good results, which can improve the quality of the solution while reducing the calculation time.

In practical applications in the field of robotics, integrated research on path planning and joint angle optimisation is increasingly being emphasised. In particular, real-time path planning and dynamic joint angle adjustment are of great importance in fields such as industrial automation and service robotics. Therefore, research based on the A* and simulated annealing algorithms is not only a theoretical discussion, but also an important requirement in practical applications [13]. The exploration of such integrated methods will help to promote the efficient operation of intelligent robots in complex environments and lay a solid foundation for future research and practice.

3 A* Algorithm and Simulated Annealing Algorithm

When planning the path for a six-degree-of-freedom robot arm in a complex environment, the A* algorithm and the simulated annealing algorithm are used together to achieve optimal path planning and joint angle optimization [14]. This comprehensive application can achieve optimization effects on both the moving path of the robot arm base and the joint angle path.

Step 1 Define the target point set and the initial node

Assume that the target point set is $T = \{t_1, t_2, ..., t_k\}$ and the initial node is S. The A* algorithm needs to plan the optimal path from the initial node S to multiple target point sets T on the grid map.

Step 2 Calculate the path cost function f(n)

The path cost function $f(n)$ of the A* algorithm consists of two parts: the actual cost $g(n)$ and the heuristic cost $h(n)$:

$$g(n) = g(p) + c(p, n) \tag{1}$$

Among them, $g(p)$ is the cost of the previous node p, and $c(p, n)$ is the cost from node p to the current node n. The cost $c(p, n)$ is usually calculated using the Euclidean distance:

$$c(p, n) = \sqrt{(x_n - x_p)^2 + (y_n - y_p)^2} \tag{2}$$

The heuristic cost h(n) is calculated by Euclidean distance:

$$h(n) = \sqrt{(x_n - x_{\text{target}})^2 + (y_n - y_{\text{target}})^2} \tag{3}$$

Among them, (x_n, y_n) is the coordinate of the current node, and $(x_{\text{target}}, y_{target})$ is the coordinate of the target node.

Step3 Path search and node expansion

Start from the starting point S and put it into the open list. For each node in the open list, select the node with the smallest $f(n)$ value to expand until the target point is reached. The expanded node is added to the open list until all target points are visited [15].

Step4 Path tracking and updating

When a node is expanded to the target node, the path is determined by backtracking its parent node and the total length of the path is calculated. If the current path reaches the target point, the path is recorded and the optimal path is updated. If the target point is not reached, continue to expand the node and repeat the above steps until all target points are visited.

Step5 Path optimization and verification

According to the final output of the A* algorithm, optimize the base movement path to ensure that the path is optimal in a complex obstacle environment. Check the rationality of the path and prepare for the next step of joint angle path optimization.

After determining the movement path of the base, the simulated annealing algorithm is introduced to further optimize the joint angle path of the robot arm. Simulated Annealing (SA) is a global optimization algorithm based on the physical annealing process [16]. By controlling the temperature to gradually decrease, the atoms in the material gradually reach the low energy state from the high energy state, thereby forming a stable crystal structure and finding the global optimal solution in a complex search space. It is particularly suitable for solving optimization problems with multiple local optimal solutions.

4 Simulated Annealing Algorithm Optimizes Joint Angle Path

Step 1 Define the Objective Function

The objective function consists of two parts: end error $E(\theta)$ and energy consumption $C(\theta)$:

$$F(\theta) = \lambda_1 E(\theta) + \lambda_2 C(\theta) \tag{4}$$

Among them, λ_1 and λ_2 are weight coefficients, and the terminal error calculation formula is:

$$(\theta) = \sqrt{(x(\theta) - x_{\text{target}})^2 + (y(\theta) - y_{\text{target}})^2 + (z(\theta) - z_{\text{target}})^2} \tag{5}$$

The energy consumption calculation formula is:

$$C(\theta) = \sum\nolimits_{i=1}^{6} \left(\frac{1}{2} I_i \omega_i^2\right) \tag{6}$$

Among them, I_i is the moment of inertia of the i-th joint, and ω_i is the average angular velocity.

Step 2 Random perturbation and solution update

Based on the current solution, a new joint angle vector θ' is generated through random perturbation, and the objective function value $F\left(\theta'\right)$ of the new solution is calculated. If $F\left(\theta'\right) < F(\theta)$, the new solution is accepted, otherwise the new solution is accepted with probability $P = \exp\left(-\frac{\Delta F}{T}\right)$, where $\Delta F = F(\theta') - F(\theta)$.

Step 3 Gradual cooling and iteration

After each iteration, the temperature T is gradually reduced according to $T(k) = T_0 \cdot \alpha^k$ until the temperature drops to the minimum set value or the maximum number of iterations is reached [17].

5 Path Visualisation and Result Analysis

In Fig. 1, a 20x20 grid map can be seen, which shows the path visualisation and the position of the grab points. The map uses different symbols to mark the five target points: Target 1 (position 1,12), Target 2 (position 8,11), Target 3 (position 16,9), Target 4 (position 16,2), and Target 5 (position 16,17). These target points are represented by symbols of different shapes, such as triangles and stars, so that the viewer can visually identify the specific location of each target.

By analysing the paths in the figure, we can observe the complexity of the movement path. The distribution of the blue areas indicates that some areas are inaccessible, forming obvious obstacles. These obstacles may affect the movement efficiency and path selection of the base. Therefore, in practical applications, the impact of these obstacles needs to be considered in order to optimise path planning and thus improve the success rate of target grasping [18]. The selection of target points and path design also need to fully consider the accessibility of space and the distribution of obstacles in the grid map, in order to achieve more effective movement and grasping strategies.

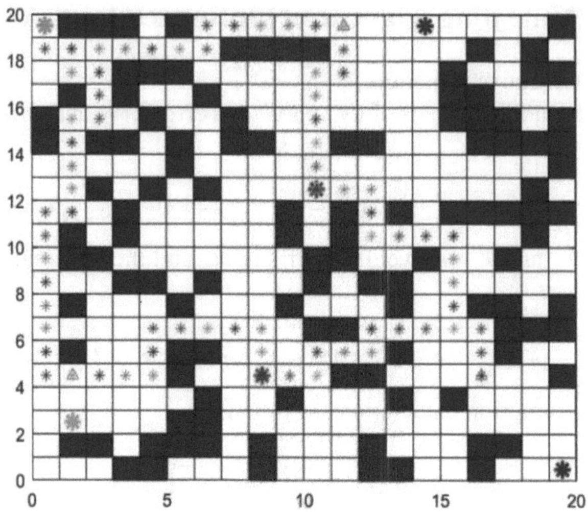

Fig. 1. Path Visualisation

In Table 1, a shortest-distance matrix is shown, containing information on the distances between the initial position and the five targets. The matrix is constructed in such a way that the distances between the individual positions are clear at a glance and provide a data basis for path planning and target grasping. Analysis of this data provides a clear picture of the relative distances between the different targets and between the targets and the initial position.

Table 1. Shortest-distance matrix

Shortest distance matrix	Initial position	Target 1	Target 2	Target 3	Target 4	Target 5
Initial position	0	16	17	25	22	38
Target 1	16	0	11	21	30	24
Target 2	17	11	0	10	19	21
Target 3	27	21	10	0	13	15
Target 4	22	30	19	13	0	28
Target 5	38	24	21	15	28	0

As can be seen from the matrix, the distances from the initial position to each goal are different. The distances to goals one and two are relatively short, [16] and [17] respectively, indicating that the paths to these two goals from the initial position are relatively direct and convenient. Goal five, however, has the longest distance, 38, indicating its relative remoteness in terms of spatial layout. This difference in distance may affect the priority of selecting the goal to be grasped in actual operation. Generally speaking, goals that are closer together are more likely to be given priority, thereby improving grasping efficiency.

In addition, the mutual distance between the targets is also worth noting. The distance between target 1 and target 2 is 11, indicating that these two targets are relatively close to each other. The path for grasping these two targets can share part of the route, which is conducive to optimising the overall movement strategy. In contrast, the distance between target 4 and target 1 is 30, which is significantly higher than that of the other targets. This may require longer time and more resources to complete the grasping task when planning the path. The shortest path is 0→4→3 →5→2→1→0, and the output end trajectory is shown in Fig. 2, with a total path of 78.

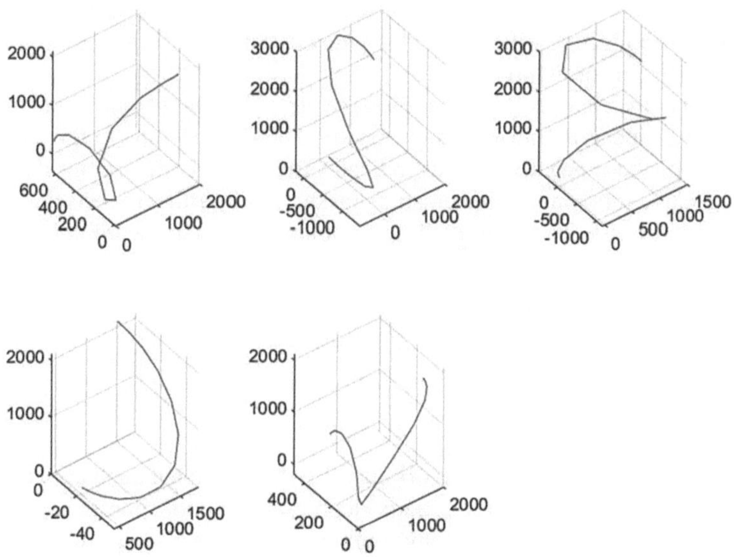

Fig. 2. End-effector trajectory

Table 2 shows the joint angles of the robot arm when grasping each target point, which were optimised by combining the A* algorithm and the simulated annealing algorithm. The setting of these angles not only directly affects the motion accuracy of the robot arm, but also plays a key role in the success rate of the grasping task. After optimisation, the minimum error of the end effector is 311.54 mm, and the minimum energy consumption is 497.82 J, which shows the effectiveness of the algorithm in path planning and energy management.

Among the data of the joint angles, the angle of the first joint changes relatively greatly, from 87.01° to 253.05°, which indicates that between different grasping points, the initial position of the robotic arm needs to be significantly adjusted to the target position. This large angle change may be because the target positions are distributed at different heights and directions in space, which causes the robotic arm to have to adjust flexibly to ensure the stability and accuracy of the grasp. The angles of the second and

Table 2. Joint angles

θ_1	θ_2	θ_3	θ_4	θ_5	θ_6
87.01°	130.63°	128.09°	1.39°	−5.24°	5.09°
110.35°	−145.15°	−233.57°	−0.25°	5.05°	−19.99°
253.05°	−149.01°	−233.55°	29.38°	−19.72°	−3.32°
3.33°	−15.28°	86.25°	−5.05°	−2.24°	7.13°
29.45°	102.91°	140.95°	0.33°	9.99°	8.05°

third joints also show large changes. In particular, the angle of the second joint between different targets changes significantly, which may be due to the high requirements on its posture during the grasping process.

In addition to the changes in joint angles, the table also shows some joint angles with negative values, such as θ_2 and θ_4, which appear to be -145.15° and -5.05°, respectively. The existence of these negative angles indicates that the robotic arm needs to perform reverse or negative rotation in some grasping tasks, which further reflects the complexity and diversity of the task. To deal with these complex path planning problems, combining the efficiency of the A* algorithm with the global optimisation ability of the simulated annealing algorithm can better adapt to the changing conditions of the robotic arm in the real environment.

6 Discussion

In this study, we combined the A* algorithm with the simulated annealing algorithm to optimise the path and joint angles of a six-degree-of-freedom robotic arm to meet the path planning challenges in complex environments. Through in-depth analysis and practical application of multiple path planning algorithms, we found that the A* algorithm has significant advantages in determining the optimal path, especially when dealing with multi-objective paths. Its heuristic search method can effectively reduce the search space of the path, enabling the robot arm to quickly find the optimal path, especially in environments with dense obstacles, which can improve movement efficiency and safety. In addition, the flexibility of the A* algorithm allows it to adapt to different environmental configurations and changes in target positions, providing the robot arm with a variety of path options.

After determining the path, we introduce the simulated annealing algorithm to further optimise the joint angles. The simulated annealing algorithm simulates the gradual cooling of a material from a high temperature to a low temperature to achieve a global optimum. This method is particularly suitable for optimisation problems with multiple local optima. In our experiments, the simulated annealing algorithm effectively reduced the error in the joint angles and energy consumption, enabling the robotic arm to reach the predetermined position with minimal energy consumption when performing grasping tasks. In this process, the setting and adjustment of parameters is particularly important, especially the choice of temperature drop and weight coefficient, which directly affect the final presentation of the optimization results.

The experimental results of this study show that the combination of the A* algorithm and the simulated annealing algorithm not only improves the efficiency of path planning, but also significantly improves the optimization of joint angles. Specifically, we observed that the minimum error and energy consumption of the end effector were both optimised, which provides a more reliable basis for the application of the robotic arm in practical operations. However, despite the positive significance of the results of this study, there are still some limitations. Future research can further explore the adaptability and real-time adjustment capabilities of these algorithms in dynamic environments, while combining with other optimisation methods, such as genetic algorithms or particle swarm optimisation, to improve overall performance.

Overall, the research on path and joint angle optimisation based on the A* and simulated annealing algorithms provides an important theoretical basis and practical guidance for the robotics field, promoting the application prospects of intelligent robots in complex environments. By continuously improving the design and application of algorithms, we expect to achieve more intelligent and adaptive robotic systems in the future, further expanding their application potential in industries, medical care, services and other fields.

7 Conclusion

This study combines the A* algorithm and simulated annealing for the path and joint angle optimization of a six-degree-of-freedom robotic arm, addressing the rising demand for intelligent robotic systems in industries such as manufacturing, healthcare, and logistics. By leveraging the A* algorithm's heuristic search capabilities, this approach enables efficient multi-objective path planning, allowing the robotic arm to navigate complex environments with obstacles while dynamically adjusting paths based on real-time environmental changes. Path visualization and cost function calculations also enhance trajectory understanding and debugging during design.

Following path determination, the simulated annealing algorithm optimizes joint angles by emulating a material's cooling process to identify global optimal solutions. Experimental results demonstrate that this combined approach reduces end-effector adjustment errors, minimizes energy consumption, and significantly improves the robotic arm's overall operational efficiency. These findings validate the feasibility of integrating A* and simulated annealing, offering valuable theoretical and practical insights for future robotic optimization studies.

However, this study recognizes the challenges posed by dynamic environmental changes, which may affect the algorithm's real-world performance. Future work could explore adaptive algorithms capable of real-time adjustment, such as integrating deep learning with optimization strategies, to enhance robotic intelligence and environmental adaptability. Furthermore, areas such as parameter tuning, algorithm integration, and real-time feedback mechanisms warrant further exploration to improve flexibility and applicability in diverse scenarios.

In conclusion, this research enriches the theoretical framework of robotic arm motion control and provides practical guidance for its application in fields like industry and medicine, offering both academic and practical value. Continued advancements in these technologies will pave the way for higher levels of intelligence and automation.

References

1. Baressi Šegota, S., Anđelić, N., Lorencin, I., Saga, M., Car, Z.: Path planning optimization of six-degree-of-freedom robotic manipulators using evolutionary algorithms. Int. J. Adv. Rob. Syst. **17**(2), 1729881420908076 (2020)
2. Jia, H., Zeng, H., Zhang, J., Wang, X., Lu, Y., Yu, L.: Robot calibration sampling data optimization method based on improved robot observability metrics and binary simulated annealing algorithm. Sensors **24**(19), 6171 (2024)
3. Xu, R., Tian, J., Li, J., Zhai, X.: Trajectory planning of rail inspection robot based on an improved penalty function simulated annealing particle swarm algorithm. Int. J. Control Autom. Syst. **21**(10), 3368–3381 (2023)
4. Yao, C., Li, Y., Ansari, M.D., Talab, M.A., Verma, A.: Optimization of industrial process parameter control using improved genetic algorithm for industrial robot. Paladyn, J. Behav. Robot. **13**(1), 67–75 (2022)
5. Jangid, M.K., Kumar, S., Singh, J.: Inverse kinematics solutions of a newly designed three-link robotic manipulator for the casting process using the ant lion optimizer. Int. J. Adv. Technol. Eng. Explor. **9**(97), 1704 (2022)
6. Kashyap, A.K., Parhi, D.R., Kumar, S.: Dynamic stabilization of NAO humanoid robot based on whole-body control with simulated annealing. Int. J. Humanoid Rob. **17**(03), 2050014 (2020)
7. Gul, F., Mir, I., Alarabiat, D., Alabool, H.M., Abualigah, L., Mir, S.: Implementation of bio-inspired hybrid algorithm with mutation operator for robotic path planning. J. Parallel Distrib. Comput. **169**, 171–184 (2022)
8. Ab Wahab, M.N., et al.: Improved genetic algorithm for mobile robot path planning in static environments. Expert Syst. Appl. **249**, 123762 (2024)
9. Pu, Q.C., Xu, X.R., Li, Q.Q., Zhang, H.: Robotic arm time–jerk optimal trajectory based on improved dingo optimization. J. Braz. Soc. Mech. Sci. Eng. **46**(4), 198 (2024)
10. Zhao, H., Zhang, B., Sun, J., Yang, L., Yu, H.: Spot-welding path planning method for the curved surface workpiece of body-in-white based on a memetic algorithm. Int. J. Adv. Manuf. Technol. **117**(9), 3083–3100 (2021)
11. Zhang, J., Yang, F., Nie, G., Zhang, J.: Optimal trajectory planning for minimizing base disturbance of a redundant space robot with IQPSO. J. Electr. Comput. Eng. **2022**(1), 3398810 (2022)
12. Luo, G., Zou, L., Wang, Z., Lv, C., Ou, J., Huang, Y.: A novel kinematic parameters calibration method for industrial robot based on Levenberg-Marquardt and differential evolution hybrid algorithm. Robot. Comput.-Integr. Manuf. **71**, 102165 (2021)
13. Liu, D.: Joint trajectory planning and simulation of a six axis welding robot. In: Proceedings of the 6th International Conference on Information Technologies and Electrical Engineering, pp. 545–552 (2023)
14. Kuang, X., Li, P., Liu, Z., Wu, C.: Inverse kinematics analysis of rock drilling manipulator based on improved SA-PSO. J. Phys.: Conf. Ser.**2450**(1), 012067 (2023)
15. Zhang, H., Li, X., Wang, L., Liu, D., Wang, S.: Construction and optimization of a collaborative harvesting system for multiple robotic arms and an end-picker in a trellised pear orchard environment. Agronomy **14**(1), 80 (2023)

16. Wang, C., Yao, X., Ding, F., Yu, Z.: A trajectory planning method for a casting sorting robotic arm based on a nature-inspired Genghis Khan shark optimized algorithm. Math. Biosci. Eng. **21**(2), 3364–3390 (2024)
17. Zhou, Y., Li, Q., Sun, Y.: Optimization of milling robot trajectory based on whale optimization algorithm. In: International Conference on Artificial Intelligence, Robotics, and Communication, pp. 337–347. Springer Nature Singapore, Singapore (2023)
18. Jangid, M.K., Kumar, S., Singh, J.: Trajectory tracking optimization and control of a three link robotic manipulator for application in casting. Int. J. Adv. Technol. Eng. Explor. **8**(83), 1255 (2021)

Stock Price Prediction Based on Investor Sentiment Analysis and CNN-LSTM-Attention Model

Maoguang Wang and Jiabei He(✉)

School of Information, Central University of Finance and Economics, Beijing 102206, China
2020312229@email.cufe.edu.cn

Abstract. Investor sentiment significantly impacts investors' decision-making behavior and stock market returns. Therefore, how to analyze and utilize investor sentiment for stock price prediction is an important research direction in finance. Many studies have explored the influence of investor sentiment on stock market price fluctuations, but these studies have not comprehensively considered factors such as overall market trends, sector rotation, and individual stock characteristics. Addressing this issue, this paper combines sentiment analysis with deep learning methods to construct a CNN-LSTM-Attention model incorporating sentiment indices, aiming to assist investors in predicting closing prices. We selected feature indicators that reflect sentiment in the overall market, sectors, and individual stock, constructed a formula for calculating sentiment indices, and calculated sentiment indices to further explore the correlation between sentiment indices and stock trends. To enhance prediction accuracy, we designed and constructed a CNN-LSTM-Attention model based on attention mechanisms, integrating convolutional neural networks and long short-term memory neural networks. This model comprehensively utilizes sentiment indices and stock data containing fundamental data, historical trading data, and technical indicators for closing price prediction. Using the stock of "iFlytek" as an example, experimental results demonstrate that compared to models such as RNN and LSTM, this model achieves the best results, indicating that incorporating sentiment indices can effectively improve prediction performance.

Keywords: Investor sentiment · Stock price prediction · Deep learning

1 Introduction

The stock market has become the primary channel for corporate fundraising and a crucial platform for stock trading. However, the complexity of stock market operations and frequent price fluctuations increase speculation opportunities and market risks. Therefore, accurate and timely stock price forecasting is essential for guiding investors to make rational decisions and maintaining the stable operation of the national economy.

However, stock prediction poses significant challenges in the field of time series forecasting. When predicting stock prices and returns, researchers often employ methods using historical data for time series analysis [1]. Traditional models in time series

X. Zhou et al. (Eds.): GPC 2024, LNCS 15225, pp. 28–42, 2026.
https://doi.org/10.1007/978-981-95-1346-8_3

analysis include Autoregressive Moving Average (ARMA) models and Autoregressive Integrated Moving Average (ARIMA) models [2]. However, due to the complexity and vast amount of information in the stock market, traditional analytical methods are inadequate, leading people to seek more efficient models to address complex time series problems. Machine learning models, with their stronger data processing capabilities, have been widely applied in financial time series analysis, such as Support Vector Machine (SVM) models [3] and Random Forest (RF) models [4]. Additionally, deep learning stands out for its excellent ability to map nonlinear relationships, enabling it to effectively address the challenges posed by the complexity of financial time series data [5, 6]. In terms of stock price prediction, deep learning has shown outstanding performance, possessing robust learning and adaptation capabilities to flexibly handle different stock market conditions and data characteristics [7]. Therefore, this study utilizes deep neural network models for stock price prediction.

Furthermore, in the stock market, investment behavior is not entirely based on rationality; the emotions and psychological characteristics of investors have a significant impact on the market [8]. A large body of research confirms that investor sentiment plays a crucial role in stock market investment decisions [9–11]. When the market is in a sustained uptrend, investors' optimistic sentiment dominates, driving them to adopt more aggressive investment strategies. Conversely, when the market experiences sustained declines, pessimistic sentiment prevails, and investors often become overly conservative, even falling into panic selling.

In existing research methods, we find that existing methods for measuring investor sentiment mostly focus on trading data in the stock market or conduct textual sentiment analysis on news reports and social media data, all limited to a single perspective. Therefore, this paper constructs a comprehensive sentiment index from three aspects: market index, sector, and individual stocks, and proposes a stock price prediction model integrating sentiment indexes. Firstly, suitable sentiment indicators are selected to construct sentiment indexes to quantify investor sentiment, and an in-depth analysis is conducted on how sentiment indexes reflect changes in investor sentiment. Secondly, a stock prediction model is constructed by integrating sentiment indexes, employing a CNN-LSTM-Attention model, and introducing multi-source information for stock price prediction. Finally, through model comparison experiments and dataset comparison experiments, the superiority of the proposed sentiment indexes and model is verified.

This paper reviews related work in Sect. 2: Sect. 3 introduces the research methodology; Sect. 4 presents our experiments and results; Sect. 5 draws conclusions.

2 Related Work

2.1 Investor Sentiment Analysis

In the academic field, the definition and measurement methods of investor sentiment have always been the focus of research. Baker and Wurgler argued that investor sentiment reflects investors' expectations and beliefs about future cash flows and risks of assets [12]. When measuring investor sentiment, three types of indicators are typically considered: objective indicators, subjective indicators, and composite indicators. Objective indicators cover factors such as closed-end fund discounts [13], IPO issuance volume and first-day returns [14], and trading volume [15]. Subjective indicators include consumer confidence indices [16] and securities analyst sentiment indices [17]. Composite indicators are constructed using methods such as principal component analysis, based on several single indicators. The BW index proposed by Baker and Wurgler is a typical example, which constructs investor sentiment factors from six observable variables [18]. Wu et al. utilized social media comments to capture investor sentiment, conducting textual sentiment analysis on post data to construct a stock investor sentiment index [19]. Thus, current research has made significant achievements in constructing sentiment indices. However, these studies still have certain limitations; they often focus on analyzing the overall market trends, historical data of individual stocks, or news commentary data, neglecting the fact that daily changes in investor sentiment of individual stocks are actually influenced by multiple factors. Therefore, this study comprehensively considers the performance of the overall market, industry sectors, and individual stocks to construct investor sentiment indices.

2.2 Stock Price Prediction

With the rapid development of artificial intelligence technology, the application of deep learning in the field of sequence data processing is becoming increasingly widespread. Recurrent Neural Networks (RNNs) have shown powerful capabilities in handling sequence data due to their unique memory mechanism, and they are highly favored by researchers [20]. However, RNNs suffer from the vanishing gradient problem, limiting their ability to capture long-term dependencies. The Long Short-Term Memory (LSTM) model proposed by Hochreiter and Schmidhuber effectively addresses this issue [21]. In addition to LSTM, Convolutional Neural Networks (CNNs) have also attracted researchers' attention due to their feature extraction capabilities. Researchers have also attempted to combine LSTM with CNNs to improve prediction accuracy. In a comparison of LSTM, CNN, and CNN-LSTM models, Aksan et al. found that hybrid deep learning models perform better than single models [22]. Vaswani et al. demonstrated that attention mechanisms can enhance the focus of time-series data prediction models on key information [23]. This mechanism has also been widely applied in prediction models, effectively improving model performance [24, 25]. Wang et al. proposed an Attention-based Long Short-Term Memory (ALSTM) model for predicting stock market price trends in four major Asian countries, demonstrating that the introduction of attention mechanisms can effectively enhance the performance of LSTM models [26]. Lu et al. proposed a CNN-BiLSTM-AM model, proving its effectiveness and accuracy in predicting stock prices [27]. Therefore, this study combines Convolutional Neural Networks and Long Short-Term Memory networks, integrating attention mechanisms, and utilizes this model to predict stock closing prices.

3 Methodology

3.1 Construction of Sentiment Index

This paper constructs comprehensive sentiment indices by building market sentiment indices, sector sentiment indices, and individual stock sentiment indices, thereby integrating them into a comprehensive sentiment index. There is a close connection between the market index, sectors, and individual stocks. The trend of the market index provides a macro background for sectors and individual stocks, the performance of sectors reflects the market sentiment of specific industries or sectors, and the specific performance of individual stocks depends on the specific situation of the companies themselves.

The comprehensive sentiment index (*CSI*) is constructed according to Eq. (1). This formula comprehensively evaluates the influence of the market sentiment index (*MSI*), sector sentiment index (*SSI*), and individual stock sentiment index (*ISI*), while considering the factors of northbound funds (*NF*) flow. These indices are weighted and combined to derive an index that comprehensively reflects investor sentiment. Northbound funds refer to the total funds buying mainland stocks through the Shanghai-Hong Kong Stock Connect and Shenzhen-Hong Kong Stock Connect. The flow of these funds often reflects judgments on the trend of the mainland stock market, with inflows considered favorable and outflows possibly related to risk aversion. The weighting factors ω_1, ω_2, and ω_3 in the formula represent the relative importance of each sentiment index in the comprehensive sentiment index, with values of 0.5, 0.3, and 0.2, respectively.

$$CSI = \omega_1 MSI + \omega_2 SSI + \omega_3 ISI + NF \tag{1}$$

The market sentiment refers to the level of market trading activity and the degree of capital inflow and outflow, reflecting the overall buying and selling sentiment of investors. The market sentiment index (*MSI*) is measured by the up-down ratio and the up-down limit ratio, as defined in Eqs. (2), (3) , and (4). The up-down ratio and the up-down limit ratio can reflect the comparative strength of the bullish and bearish forces in the market, and are used to judge the trend of market rises and falls. Here, S_a represents the number of stocks that went up in the A-share market on a certain day, S_d represents the number of stocks that went down, N_{up} represents the number of stocks that hit the up limit, and N_{down} represents the number of stocks that hit the down limit. *ADR* represents the up-down ratio, *UDR* represents the up-down limit ratio, and α_1 and α_2 represent the respective weights of 0.6 and 0.4 for the up-down ratio and up-down limit ratio.

$$MSI = \alpha_1 ADR + \alpha_2 UDR \tag{2}$$

$$ADR = \frac{S_a}{S_d} \tag{3}$$

$$UDR = \frac{N_{\text{up}}+1}{N_{\text{down}}+1} \tag{4}$$

Sector sentiment refers to the collective sentiment and expectations of investors towards a specific sector. The Sector Sentiment Index (*SSI*) is measured by the price changes of leading stocks in the sector and the price changes of the Shanghai and Shenzhen Stock Exchange Industry Index, defined as Eq. (5). The price changes of leading stocks in the sector effectively reflect the overall market direction of the sector, guiding and demonstrating the rise and fall of individual stocks within the same sector. The price changes of the Shanghai and Shenzhen Stock Exchange Industry Index are closely related to the performance of stocks within the sector. Observing their changes can capture industry and market trends, aiding in the analysis of sector trends. Here,

LTI represents the price change of leading stocks in the sector, *SWI* represents the price change of the Shanghai and Shenzhen Stock Exchange Industry Index, and β_1 and β_2 represent the weights of the price changes of leading stocks in the sector and the Shanghai and Shenzhen Stock Exchange Industry Index, respectively, with values of 0.7 and 0.3.

$$SSI = \beta_1 LTI + \beta_2 SWI \tag{5}$$

Individual stock sentiment typically refers to the sentiment and attitude of investors towards specific stocks or companies, manifested as optimistic, pessimistic, or neutral. This paper constructs a sentiment index and optimizes it to be more suitable for the current application scenario, termed as Individual Stock Sentiment Index (*ISI*), as shown in Eqs. (6) and (7).

$$ISI = ln\left[\frac{1+N_{pos}}{1+N_{neg}}\right] * PI \tag{6}$$

$$PI = \ln(1 + Read + Comment + Thumbup) \tag{7}$$

In which N_{pos} and N_{neg} represent the number of daily bullish and bearish posts respectively, *Read* indicates the average number of post reads, *Comment* indicates the average number of post reads, *Thumbup* signifies the average number of post likes, and *PI* stands for the post's influence. Specifically, the number of reads reflects the frequency of post views, the number of comments reveals the level of user engagement, and the number of likes reflects the level of user approval towards the post content.

3.2 CNN-LSTM-Attention Neural Network Model

This study constructs a CNN-LSTM-Attention neural network model, whose architecture is illustrated in Fig. 1. The model consists of three key components: Convolutional Neural Network (CNN), Long Short-Term Memory (LSTM) network, and Attention mechanism. Firstly, 1D-CNN is utilized to extract data features. After passing through the convolutional layers, the data undergoes nonlinear transformation via the ReLU activation function, followed by dimensionality reduction through the max-pooling layer. Subsequently, LSTM is employed to handle time series data while incorporating the Attention mechanism. The incorporation of this mechanism enhances the model's focus on key parts of the input data, effectively improving the prediction accuracy of the model. Next, further feature extraction is performed through fully connected layers, and dropout techniques are applied to reduce the risk of overfitting. Finally, the output layer produces the final prediction results. This model combines the excellent feature extraction capabilities of CNN, the ability of LSTM to handle time series data, and the attention mechanism's ability to focus on important information, thereby providing the model with advantages in handling time series data.

Firstly, feature extraction of the input data is performed using a one-dimensional Convolutional Neural Network (1D-CNN) consisting of convolutional layers and pooling layers. The 1D-CNN slides convolutional kernels along the time steps of the input sequence, computing the dot product between the convolutional kernel and each window of the input sequence, thereby generating new feature representations. The input to this

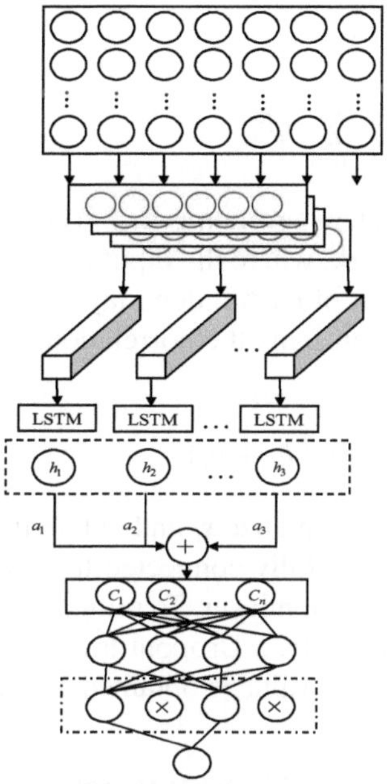

Fig. 1. CNN-LSTM-Attention model structure schematic diagram

network is a time series, starting from a time point, obtaining a sequence composed of continuous feature vectors. The calculation process of the convolutional layer can be described using Eq. (8).

$$k_i^l = f\left(b^{(l)} + \sum_{l=0}^{len} W_c^{(l)} \cdot x_{i+l}^{(l)}\right) \tag{8}$$

where $k_i^{(l)}$ is the output result of the k-th layer convolutional kernel in the i-th channel f is the activation function, and the ReLU function is used. *len* is the length of the convolutional kernel, $b^{(l)}$ is the bias term, $W_c^{(l)}$ is the weight matrix, and $x_{i+l}^{(l)}$ is the output of the previous layer convolutional layer.

Then, further reduction of the feature map size outputted by the convolutional layers is achieved through pooling layers, effectively reducing data dimensionality while retaining crucial local features within the time series. This study utilizes the max-pooling function. Following the processing by the convolutional layers and pooling layers, the data is inputted into the LSTM layer to extract temporal features.

After the LSTM layer, an Attention layer is added to enable the model to focus on key information. The hidden layer vectors outputted by the LSTM layer undergo weighted summation through the Attention layer. During this process, the importance of features at each time step is represented by corresponding weights. Assuming the input is m feature vectors h_i, the model obtains the context vector c_i based on these vectors, as shown in Eq. (9):

$$c_i = \sum_{i=1}^{k} a_i h_i \tag{9}$$

where the weights a_i at different time steps are computed. The computation of a_i is achieved through a fully connected network, where the input is the hidden layer vectors outputted by the LSTM layer. The impact of each hidden layer vector on the output is evaluated by calculating a score s_i. s_i represents the correlation between h_i and c_i, as shown in Eq. (10):

$$s_i = \tanh\left(w^T h_i + b_i\right) \tag{10}$$

The softmax function is employed to normalize s_i and obtain the final weight coefficients a_i. Following the Attention layer, a fully connected layer is utilized to further extract features from the output of the Attention layer. To mitigate the risk of overfitting and enhance the model's generalization ability, a Dropout layer is introduced after the fully connected layer. Finally, the prediction value is obtained as the output.

3.3 Constructing a Stock Price Prediction Model Integrating Sentiment Index

This section introduces a stock price prediction model that integrates sentiment index, as illustrated in Fig. 2. Firstly, the investor sentiment index is computed, and its relationship with market sentiment changes is analyzed. Then, correlation analysis is conducted to select features. Utilizing the CNN-LSTM-Attention model, which combines spatial feature extraction of CNN, temporal sequence analysis of LSTM, and attention mechanism, the model captures the impact of sentiment on stock prices. By adjusting hyperparameters to optimize the model, the best performance is achieved on the validation set. The predictive performance of the model is quantified using a series of evaluation metrics. After thorough training and evaluation, the resulting model effectively predicts the closing price of stocks for the next trading day.

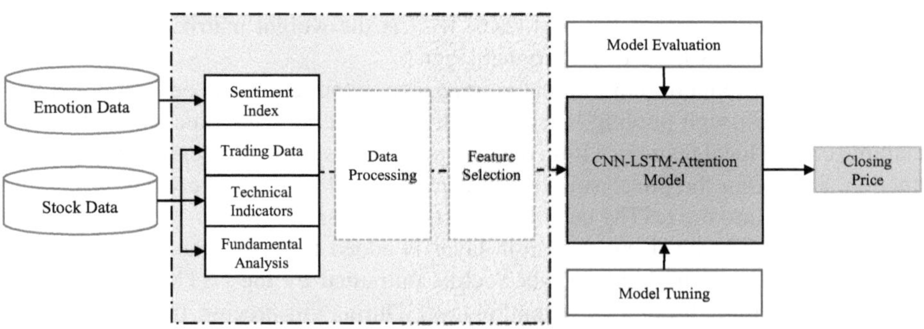

Fig. 2. Model Architecture Diagram

4 Experiments

4.1 Data Description

In the experiment, the stock of "iFlytek" listed on the Shenzhen Stock Exchange is selected as the research subject. The data sample spans from August 1, 2018, to October 15, 2021, covering a total of 777 trading days of daily data. The experimental dataset is primarily divided into two parts: sentiment data and stock data. Section 3.1 provides a detailed description of the composition of the sentiment dataset. The stock data includes trading data, technical indicators, and fundamental data, as shown in Table 1.

Table 1. Stock data in detail

Stock data	Specific data
Trading data	Opening price, closing price, highest price, lowest price, trading volume, turnover, amplitude, price change, percentage change, turnover rate
Technical indicators	Exponential Moving Average (EMA), Relative Strength Index (RSI)
Technical indicators	Price-to-Earnings Ratio, Price-to-Book Ratio, Price-to-Sales Ratio

4.2 Calculation and Analysis of Sentiment Index

In the experiment, based on the sentiment index formula constructed in Sect. 3.1, the daily sentiment index is calculated. Firstly, missing data is filled using linear interpolation, and then the data is processed. The logarithmic function is applied to handle the ratio of rising to falling stocks, while Z-Score normalization is performed on the number of stocks hitting the daily limit. Preprocessing is conducted on the net inflow of northbound funds data, where if the net inflow of northbound funds on a certain day is greater than zero, it is divided by the maximum positive value; if it is less than zero, it is divided by the absolute value of the minimum negative value. The sentiment index from August 1, 2018, to October 15, 2021, is calculated, and it is smoothed using the moving average method. Based on this method, a trend chart of the closing price and sentiment index is plotted, as shown in Fig. 3. From the chart, it can be observed that there is a certain correlation between the sentiment index and the stock price trend. When the sentiment index is positive, it typically corresponds to a bullish market sentiment, and the stock price trend is relatively positive, indicating investor sentiment is optimistic. Conversely, when the sentiment index is negative, the market sentiment and stock price trend often decline synchronously, indicating investor sentiment is pessimistic.

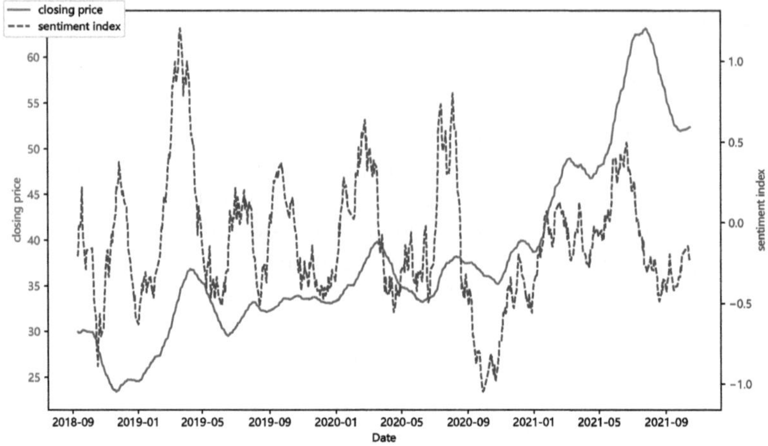

Fig. 3. Sentiment Index and Stock Price Trends

Further observation reveals that the inflection points of sentiment indices are close to those of closing prices, suggesting that when investor sentiment undergoes extreme changes, market prices often experience phase reversals and may begin to fluctuate in the opposite direction. This implies that the psychological state and behavior of market participants directly influence market prices, and price movements swiftly feedback and affect sentiment, creating cyclical fluctuations. This indicates a close correlation between the sentiment index and stock price trends, which is helpful for short-term stock price predictions.

To explore the relationship between sentiment indices and investor sentiment, as well as their specific impact on stock price trends, this experiment selected extreme points of sentiment indices for analysis. Next, we will present the analysis results when the sentiment index reaches its maximum value.

On February 25, 2019, the sentiment index reached a peak value of 6.515, reflecting extreme optimism in the market sentiment. Within the 7 days before and after this time point, data on sentiment index, closing price, price change rate, investor sentiment, and stock price trends are shown in Table 2 and Fig. 4. In the graph, a sentiment index greater than 0 represents optimistic market sentiment, while negative values represent pessimism. The analysis is as follows:

- 2019/2/20 to 2019/2/21: The sentiment index remained in the negative territory, with values of -1.321 and -1.041 respectively, reflecting widespread pessimism among market investors. This pessimistic sentiment is consistent with the downward trend in stock prices.
- 2019/2/22: The sentiment index experienced a reversal, changing from a negative value of -1.859 to a positive value. This transformation aligns with the upward trend in stock prices. It may imply that the positive shift in market sentiment has a positive driving effect on stock prices.
- 2019/2/25: The sentiment index surged to 6.515, reflecting a significant improvement in market sentiment. Driven by this optimistic sentiment, the stock price experienced increases of 8.02% and 8.81% respectively in the following two trading days.

- 2019/2/26 to 2019/2/27: After February 25th, market sentiment remained elevated, although the sentiment index experienced a slight decrease, it remained positive. The market responded positively, with stock prices showing strong gains, reaching an increase of 8.81% on February 26th. This validates the positive correlation between sentiment index and stock prices, as well as the market's continued response to positive sentiment.
- 2019/2/28: The sentiment index declined to 0.224, indicating a market adjustment to the recent uptrend. Subsequently, stock prices fell by 2.86%. This shift in sentiment may prompt investors to adopt more cautious trading strategies to cope with market fluctuations.

Table 2. Analysis of Emotion and Stock Price Trends from February 20, 2019, to February 28, 2019

Date	Sentiment index	Closing price	Change	Investor sentiment	Price Trend
2/20	−1.321	30.37	−1.3	Pessimistic	Decrease
2/21	−1.041	30.04	−1.09	Pessimistic	Decrease
2/22	1.859	30.69	2.16	Optimistic	Increase
2/25	6.515	33.15	8.02	Optimistic	Increase
2/26	1.192	36.07	8.81	Optimistic	Increase
2/27	1.335	36.76	1.91	Optimistic	Increase
2/28	0.224	35.71	−2.86	Optimistic	Decrease

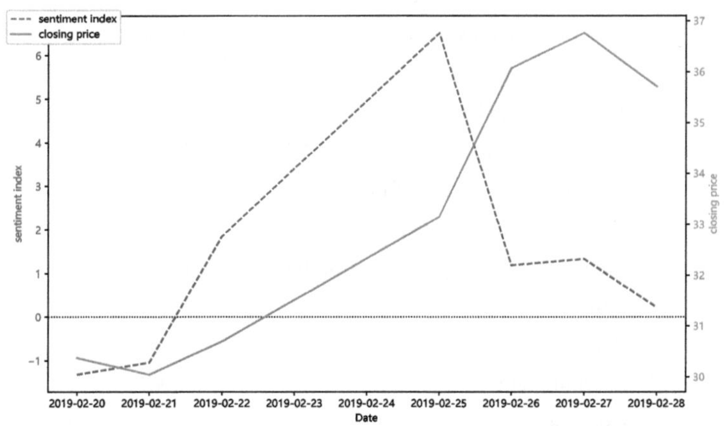

Fig. 4. Sentiment Index and Closing Price Trends from February 20, 2019, to February 28, 2019

From the above analysis, it can be concluded that the positive or negative values of the emotion index correspond to the market sentiment being optimistic or pessimistic. When the emotion index is positive, it indicates a tendency for the stock market to rise in the future; whereas when the emotion index is negative, it implies that the market may face the risk of decline. Additionally, the magnitude of the emotion index is a key indicator of the intensity of sentiment. The larger the absolute value, the more intense the market sentiment, and correspondingly, the greater the magnitude of the stock price rise or fall. The changes in the emotion index exhibit long-term correlations with stock price trends, with turning points often synchronizing with key changes in stock prices.

4.3 Model Implementation and Evaluation Metrics

We combined the calculated sentiment indices with stock data and conducted a correlation analysis on these features. To optimize the model, we removed features with correlation coefficients exceeding 90% to reduce redundancy. Ultimately, we identified a set of key input features, including closing price, trading volume, price change rate, turnover rate, Exponential Moving Average (EMA), Relative Strength Index (RSI), Price-to-Earnings Ratio, and Price-to-Book Ratio. These features will be used for subsequent model training and prediction. Through a combination of random search and grid search for hyperparameter tuning experiments, the parameters of the model were determined as shown in Table 3.

Table 3. Parameters of the CNN-LSTM-Attention Model

Parameter	Value
Convolutional layer filters	64
Convolutional layer kernel size	3
Convolutional layer activation function	RELU
Convolutional layer padding	Same
Pooling layer pool size	2
Number of LSTM layers	1
Number of hidden units in LSTM layer	512
LSTM layer activation function	Tanh
Number of fully connected layer	2
Fully connected layer activation function	RELU
Time step	3
Batch size	16
Learning rate	0.001
Dropout	0.5
Epochs	100
Optimizer	Adam

This paper selects fundamental evaluation metrics for regression models, including Mean Squared Error (MSE), Root Mean Squared Error (RMSE), Mean Absolute Error

(MAE), Mean Absolute Percentage Error (MAPE), and Coefficient of Determination (R-squared).

4.4 Experimental Results and Performance Comparison

To validate the superiority of the CNN-LSTM-Attention model, we compared its performance on the test set with that of other models, including RNN, LSTM, Gated Recurrent Unit (GRU), LSTM with Attention Mechanism (LSTM-AM), CNN, and the fusion of Convolutional and Long Short-Term Memory networks (CNN-LSTM), all based on the dataset incorporating sentiment indices. The experimental results of the model comparisons are shown in Table 4, where the model proposed in this paper is referred to as CNN-LSTM-AM. All data in the table are predictions made using the dataset containing sentiment indices.

Table 4. Experimental Results of Model Comparisons

Model	MSE	RMSE	MAE	MAPE	R^2
RNN	21.3652	4.6222	3.4764	0.0572	0.1906
GRU	16.2380	4.0296	3.3681	0.0567	0.3849
LSTM	14.5384	3.8129	3.0276	0.0505	0.4493
LSTM-AM	12.6109	3.5512	2.7538	0.0458	0.5223
CNN	4.6251	2.1506	1.8378	0.0323	0.8248
CNN-LSTM	3.5133	1.8744	1.3867	0.0235	0.8669
CNN-LSTM-AM	**2.4258**	**1.5575**	**1.1474**	**0.0196**	**0.9081**

Experimental results show that the basic RNN model performs poorly, with high MSE values and large prediction errors. GRU and LSTM models show improvement in error metrics through their complex gating mechanisms. The LSTM-AM model further enhances prediction accuracy by introducing an attention mechanism. CNN models and CNN-LSTM models, with their superior feature extraction capabilities, achieve lower error metrics compared to earlier models, demonstrating good performance on time series data. Our proposed CNN-LSTM-AM model exhibits excellent performance across all evaluation metrics, with the lowest MSE value and R^2 score exceeding 0.9, indicating its highest predictive accuracy. In conclusion, the CNN-LSTM-AM model proposed in this study demonstrates good performance in stock prediction.

To validate the effectiveness of sentiment index in stock prediction, we compared the performance of models with and without sentiment index features. The dataset Stock contains only stock data, while Emotion + Stock includes sentiment index data. This comparative experiment employs the CNN-LSTM-Attention model for prediction.

Table 5. Comparative Results of Datasets

Dataset	MSE	RMSE	MAE	MAPE	R^2
Stock	4.2995	2.0735	1.5744	0.0265	0.8371
Emotion + Stock	**2.4258**	**1.5575**	**1.1474**	**0.0196**	**0.9081**

The experimental results, as shown in Table 5, reveal that the model incorporating sentiment index features outperforms the model without sentiment index in all evaluation metrics, indicating that sentiment index can enhance the performance of stock price prediction models. As illustrated in Fig. 5, the model without sentiment index exhibits a considerable gap between predicted values and actual values, while the model with sentiment index can predict more accurately, with predicted values closer to the actual values, thereby demonstrating higher prediction accuracy.

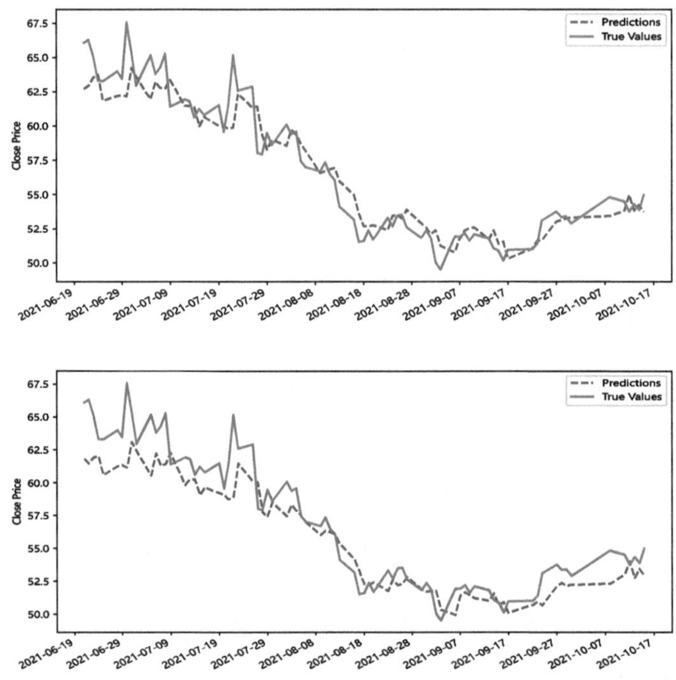

Fig. 5. Stock Price Prediction Graphs based on datasets Stock (top) and Emotion + Stock (bottom)

5 Conclusion

This study investigates stock market investment prediction based on investor sentiment analysis. The sentiment index constructed in this research exhibits a clear correlation with stock price trends. The sentiment index also reflects the intensity of emotions, where a larger absolute value indicates a more concentrated and unified market sentiment.

Moreover, the turning points of the sentiment index correspond closely to the inflection points of stock prices, indicating the significant impact of extreme changes in investor sentiment on market prices. The CNN-LSTM-Attention model developed in this paper demonstrates promising results in stock price prediction. Additionally, we validate the effectiveness of sentiment indices in stock prediction.

Currently, existing sentiment indicators do not cover all factors influencing market sentiment. In the future, it is necessary to consider changes in macroeconomic policies, the impact of major news events, and changes in political situations, among other factors, in order to construct a more comprehensive sentiment index system. This will more accurately reflect changes in market participants' sentiment and provide a more reliable basis for predicting market trends. Furthermore, we plan to utilize Transformer-based technologies to analyze sentiment and news sentiment, aiming to further enhance the predictive effectiveness of the model.

Disclosure of Interests. The authors have no competing interests to declare that are relevant to the content of this article.

References

1. Efendi, R., Arbaiy, N., Deris, M.M.: A new procedure in stock market forecasting based on fuzzy random auto-regression time series model. Inf. Sci. **441**, 113–132 (2018)
2. Idrees, S.M., Alam, M.A., Agarwal, P.: A prediction approach for stock market volatility based on time series data. IEEE Access **7**, 17287–17298 (2019)
3. Li, X., Sun, Y.: Stock intelligent investment strategy based on support vector machine parameter optimization algorithm. Neural Comput. Appl. **32**(6), 1765–1775 (2020)
4. Park, H.J., Kim, Y., Kim, H.Y.: Stock market forecasting using a multi-task approach integrating long short-term memory and the random forest framework. Appl. Soft Comput. **114**, 108106 (2022)
5. Yang, L., et al.: NumHTML: numeric-oriented hierarchical transformer model for multi-task financial forecasting. In: AAAI Conference on Artificial Intelligence, vol. 36, no. 10, pp. 11604–11612 (2022)
6. Wei, T.X., You, Y.N., Chen, T.L.: AR-stock: deep augmented relational stock prediction. In: The AAAI-21 Workshop on Knowledge Discovery from Unstructured Data in Financial Services (2021)
7. Wang, H., et al.: Adaptive long-short pattern transformer for stock investment selection. In: IJCAI, pp. 3970–3977 (2022)
8. Gross, J.J.: Emotion regulation: taking stock and moving forward. Emotion **13**(3), 359 (2013)
9. Griffith, J., Najand, M., Shen, J.: Emotions in the stock market. J. Behav. Financ. **21**(1), 42–56 (2020)
10. Lazzini, A., et al.: Emotions, moods and hyperreality: social media and the stock market during the first phase of COVID-19 pandemic. Acc. Audit. Account. J. **35**(1), 199–215 (2022)
11. Goodell, J.W., et al.: Emotions and stock market anomalies: a systematic review. J. Behav. Exp. Financ. **37**, 100722 (2023)
12. Baker, M., Wurgler, J.: Investor sentiment in the stock market. J. Econ. Perspect. **21**(2), 129–151 (2007)
13. Hudson, Y., Yan, M., Zhang, D.: Herd behaviour & investor sentiment: evidence from UK mutual funds. Int. Rev. Financ. Anal. **71**, 101494 (2020)
14. Liu, X., Wu, B.: Do IPO firms misclassify expenses? Implications for IPO price formation and post-IPO stock performance. Manag. Sci. **67**(7), 4505–4531 (2021)

15. Duz Tan, S., Tas, O.: Social media sentiment in international stock returns and trading activity. J. Behav. Finance **22**(2), 221–234 (2021)
16. Malovaná, S., Hodula, M., Frait, J.: What does really drive consumer confidence? Soc. Indic. Res. **155**(3), 885–913 (2021)
17. Cuculiza, C., Antoniou, C., Kumar, A., Maligkris, A.: Terrorist attacks, analyst sentiment, and earnings forecasts. Manag. Sci. **67**(4), 2579–2608 (2021)
18. Baker, M., Wurgler, J.: Investor sentiment and the cross-section of stock returns. J. Finance **61**(4), 1645–1680 (2006)
19. Wu, S., Liu, Y., Zou, Z., Weng, T.H.: S_I_LSTM: stock price prediction based on multiple data sources and sentiment analysis. Connect. Sci. **34**(1), 44–62 (2022)
20. Hewamalage, H., Bergmeir, C., Bandara, K.: Recurrent neural networks for time series forecasting: current status and future directions. Int. J. Forecast. **37**(1), 388–427 (2021)
21. Hochreiter, S., Schmidhuber, J.: Long short-term memory. Neural Comput. **9**(8), 1735–1780 (1997)
22. Aksan, F., Li, Y., Suresh, V., Janik, P.: CNN-LSTM vs. LSTM-CNN to predict power flow direction: a case study of the high-voltage subnet of northeast Germany. Sensors **23**(2), 901 (2023)
23. Vaswani, A., et al.: Attention is all you need. In: Advances in Neural Information Processing Systems, vol. 30 (2017)
24. Lin, Z., Li, M., Zheng, Z., Cheng, Y., Yuan, C.: Self-attention ConvLSTM for spatiotemporal prediction. In: AAAI Conference on Artificial Intelligence, vol. 34, no. 07, pp. 11531–11538 (2020)
25. Jiang, J., Han, C., Zhao, W.X., Wang, J.: Pdformer: Propagation delay-aware dynamic long-range transformer for traffic flow prediction. In: AAAI Conference on Artificial Intelligence, vol. 37, no. 4, pp. 4365–4373 (2023)
26. Wang, J., Cui, Q., Sun, X., He, M.: Asian stock markets closing index forecast based on secondary decomposition, multi-factor analysis and attention-based LSTM model. Eng. Appl. Artif. Intell. **113**, 104908 (2022)
27. Lu, W., Li, J., Wang, J., Qin, L.: A CNN-BiLSTM-AM method for stock price prediction. Neural Comput. Appl. **33**(10), 4741–4753 (2021)

Losslessly Selective Chaotic Encryption for Medical Images Based on U^2-Net and Stegonography

Xianhua Song(✉), Ying Liu, and Di Jiang

School of Science, Harbin University of Science and Technology, Harbin 150080, China
songxianhua@hrbust.edu.cn

Abstract. Massive high-resolution medical images have become an important medium for medical record storage, disease diagnosis, online consultation and remote diagnosis and treatment to ensure people's health. Traditional full-image encryption technology is an important means of safeguarding patient privacy. However, for massive high-resolution medical images, traditional encryption technology cannot meet the requirements for high processing speed. Due to the significant separation between organ regions and background regions in medical images, local encryption technology for regions of interest (ROI) in medical images is expected to meet the high processing rate requirements for large volumes of medical images. Meantime, the non-destructiveness of decrypted medical images is necessary to ensure the precision medicine. This paper primarily studies the efficient, secure, and lossless encryption and decryption technology for the local ROI in medical images. The binary edge image of the medical image is obtained using U^2-Net, thus the original image is divided into ROI and the region of the background (ROB) according to the positional relationship between the original image and the ROI in the binary image. The ROI bit plane is scrambled using the Chen chaotic system, and the key matrix is generated by the Henon chaotic map to diffuse the scrambled bit plane image. The encrypted ROI and the unencrypted ROB are combined into a ciphertext image. Run-length encoding is used to compress and encode the ROI binary image, then the encoding results are randomly embedded into the ciphertext image, which can achieve lossless decryption while improving security.

Keywords: Medical Images · Selective Encryption · U^2-Net · Chaos System · Run-Length Encoding

1 Introduction

While the convenience of information sharing facilitates people's lives, it also raises the issue of information security protection [1]. The digital informatization of medical data also has multiple information security risks [2]. Medical images as a kind of important medical information, contain sensitive information that, if maliciously intercepted will face risks such as leakage of patients' private information or illegal theft. Medical image encryption is crucial to ensure the security of medical images and prevent medical information from being leaked [3].

X. Zhou et al. (Eds.): GPC 2024, LNCS 15225, pp. 43–65, 2026.
https://doi.org/10.1007/978-981-95-1346-8_4

With the acceleration of the process of medical informatization, massive medical image data show exponential growth. In the face of increasing medical image data, how to achieve lossless encryption and decryption while ensuring efficient processing has become a major challenge for current research. Traditional encryption algorithms are difficult to meet the real-time demand due to high computational complexity and long processing time; while lightweight encryption methods improve the processing efficiency, but may lead to data loss due to compression and other processes, while any pixel distortion in medical diagnosis may trigger the risk of misdiagnosis.

Due to the unique nature of medical images, such as the high resolution of medical images and the significant separation of organ and background regions, traditional whole-image-based encryption techniques may not be able to guarantee the timeliness of encryption. Therefore, there is a need to develop enhanced and specialized selective encryption algorithms for local organs of medical images.

Image selective encryption is generally divided into two steps: ROI region extraction and ROI region encryption. The earliest ROI region extraction algorithms were mainly proposed based on people's a priori thinking. The earliest proposed ROI extraction algorithm is based on the design of contrast features, i.e., the region with significant contrast with the background is the ROI. Itti et al. [4] calculated the contrast of "Center-Surround" based on the visual effect of primates through the dimensions of orientation, color and contrast features. Cheng et al. [5] realized ROI extraction by considering the consistency of a part of the image with the surrounding pixels and the contrast of that part with the global pixels in parallel. Dinesh Pandey [6] proposed a segmentation method for magnetic resonance imaging images. Andrey Gaide [7] proposed a scheme for automatic extraction of ROIs for images or lungs e-computed tomography.

With the development of technology, people propose to design algorithms for ROI extraction based on feature learning, i.e., using machine learning algorithms to continuously learn the features in the image and then derive the ROI region that people need. The main representatives are Vig et al. [8] designed the Ensemble of Deep Networks (EDN) model, which extracts the features by neural networks and implements the decision-making from features to ROIs according to Support Vector Machines (SVM). Liu et al. [9] proposed Mr-CNN model which uses convolutional neural network to obtain multi-scale features of fully connected layers and pixels and then continuously train the salient values of the pixels. Kruthiventi et al. [10] proposed Deep Fx model which trains and predicts the multibranch features. Pan et al. [11] designed respectively a shallow and a deep end-to-end learning neural network model for in turn extracting ROI.

For image ROI region encryption, G. Bhatnagar et al. [12] proposed an image selective encryption algorithm using pixels of interest (POI) and singular value decomposition, this encryption scheme uses sawtooth filled curve algorithm to scramble the pixels, and then achieves the diffusion operation on the ROIs by using the nonlinear chaotic mapping and singular value decomposition. Wu et al. [13] proposed a one-dimensional Wheel-Switch chaotic system and combined the scrambling and permutations framework to encrypt images. Chen et al. [14] implemented a scrambling method with respect to the adjacent pixels using a scrambling-diffusion encryption framework. Wang et al. [15] realized image block encryption using dynamic random growth method. Xu et al.

[16] combined block scrambling and dynamic index diffusion algorithms to encrypt images.

After the security and encryption effectiveness of image encryption are guaranteed, researchers have shifted their goal to encryption efficiency. Due to the high security and fast encryption efficiency of bit plane, it is widely used to encrypt images. Mozaffari et al. [17] proposed the use of local binary and bit plane decomposition methods. Som et al. [18] decompose the image into eight bit planes, then classify the bit planes according to the threshold value, and then encrypt only some of the effective bit planes and reorganize the eight bit planes into encrypted image. Yin et al. [19] designed Reversible Data Hiding in Encrypted Images (RDHEI) encryption algorithm using pixel prediction and bit plane compression.

Main contributions are summarized:

(1) Application of U^2-Net to the field of ROI encryption of head medical images ensures accuracy and completeness of ROI extraction.
(2) Using chaotic system to encrypt part of the bit planes of ROI of medical images, which effectively improves the security of medical images under the premise of guaranteeing encryption efficiency.
(3) Apply run-length encoding to convert the binary image into a bit stream and embed it into the encrypted image, thus further ensuring the security of the ciphertext image and achieving ROI lossless decryption.

2 Related Work

In order to design a lossless selective encryption scheme for medical images that is both secure and efficient, this chapter briefly introduces the knowledge of medical image ROI extraction, encryption and lossless decryption. It mainly includes U^2-Net, Chen chaotic system, Henon chaotic system and run-length encoding.

2.1 U^2-Net Detection

U^2-Net is actually nested by two-layer U-Net structure. U-Net evolved from a fully convolutional neural network, which was named U-Net [20] because of its resemblance to the letter U in its network structure. The network consists of two main paths, search and expansion. The search path serves to obtain the contextual content in the image and the expansion path can pinpoint to the target region in the image. U^2-Net is mainly used in the field of Salient Object Detection (SOD). It separates the target region from the background region in an image, allowing the image to be divided into foreground and background. Therefore, the application of U^2-Net to the medical image field is more in line with the application background.

U^2-Net has two advantages: (1) The mixture of various degrees of sensory fields in the Residual U-block (RSU) enables access to rich contextual content from several scales; (2) A pooling operation is performed in the RSU-block, which adds deeper architectural depth but does not significantly increase computational cost [21].

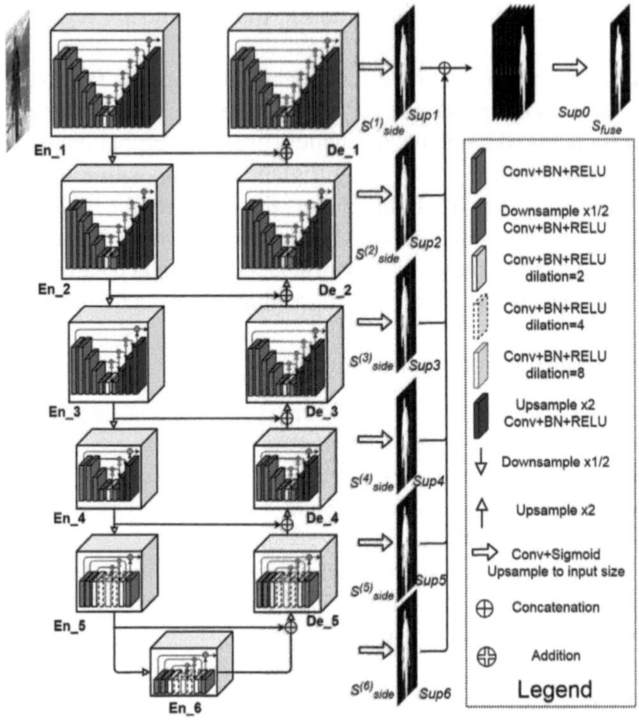

Fig. 1. Architecture of U^2-Net [21]

U^2-Net consists of three main parts, the architecture of U^2-Net is schematically shown in Fig. 1:

(1) Six-level encoder: In encoder levels 1–4, RSU blocks with heights (L) of 7, 6, 5 and 4 are used, respectively. Typically, L is assigned as the spatial resolution of the feature map. For higher resolution images larger L is needed to cover more content. For lower resolution feature maps like En_5 and En_6, implementation of downsampling may make important information missing. Therefore, RSU-4F ("F" is the promoted application of RSU) is generally applied at this stage. The application of RSU-4F can make the resolution of the originally lower-resolution feature maps consistent with the resolution of the input feature maps.
(2) Five-stage decoder: The decoder has a similar structure to the encoder En_6. The same expansion RSU block is used in decoder De_5. The input of each decoder is the connection of the previous upsampling feature map and the corresponding coding feature map.
(3) Saliency Map Fusion: Similar to HED [22], first U^2-Net generates six saliency probability maps S(6) – S(1) based on the encoder En_6 and the decoder De_5-De_1 through the convolutional layer of 3 × 3 and the Sigmoid function. Then, these saliency maps are upsampled to the input image size and fused together by the Concat operation. Finally, the significance probability maps S are derived by 1 × 1 convolutional layers and Sigmoid function.

The method detects more accurate medical image ROIs with smoother edges, which can be extracted more effectively.

2.2 Chen Chaotic System

Chen chaotic system was proposed by Prof Guanrong Chen, which is similar to the Lorenz chaotic system, but it is slightly more complex in phase space. [23]. There are many algorithms for image encryption using Chen chaotic system [24, 25], which is mathematically modelled as Eq. (1):

$$\begin{cases} \frac{dx}{dt} = a(y - x) \\ \frac{dy}{dt} = (c - a)x - xz - cy \\ \frac{dz}{dt} = xy - bz \end{cases} \tag{1}$$

where x, y and z are the state quantities, $a > 0$, $b > 0$, $c > 0$ are the parameters. Also, the system can exhibit a chaotic state when $a = 35$, $b = 3$, and $c = 28$.

Chen chaotic system is not topologically equivalent, while the system is less predictable. Therefore, it is feasible to apply this system to encrypt images.

2.3 Henon Chaotic Map

The Henon chaotic map is a discrete time dynamical system that can generate chaotic phenomena [26, 27], which is mathematically modelled as Eq. (2):

$$\begin{cases} s_{n+1} = 1 + t_n - ps_n^2 \\ t_{n+1} = qs_n \end{cases} \tag{2}$$

where s and t are the input parameters and n is iteration times. When the control parameters $p = 1.4$ and $q = 0.3$, the system is in a chaotic state. This state can generate chaotic sequences with high randomness.

2.4 Run-Length Encoding

Run-length encoding (RLE) is a relatively simple form of data compression. The run of data (i.e., the sequence of consecutive occurrences of the same data in a sequence of data) is stored as the current value and the number of occurrences is counted [28].

The pixel values in a binary image are only "0" and "1", and thus appear consecutively in the image, usually the section with "0" is called the "0 "tour, the same will be even" 1 "this section is called" 1" tour. For example, assuming that the sequence 000111011011... is known, then according to the encoding rules of the RLE, we can obtain the run-length sequence 331212....

The RLE encoding method is simple and extremely fast, so it can be used to speed up decompression. Decompression of binary images using RLE enables lossless decryption and reduces runtime.

3 Medical Image ROI Encryption and Decryption Scheme

A high-security medical image ROI encryption algorithm based on bit plane and controllable compression is proposed, which can not only improve security but also ensure efficiency. The encryption algorithm consumes part of the running time when extracting ROIs, but improves the accuracy and completeness of medical image ROI extraction, thus improving the security of medical image ROIs. This section describes the specific encryption and decryption algorithm.

3.1 Encryption Scheme

Figure 2 depicts the basic architecture of ROI extraction and ROI encryption using bit plane and controlled compression. The encryption scheme employs U^2-Net to extract ROIs, and then partially bit-plane encrypts the ROIs to balance the whole runtime. This method not only ensures the encryption efficiency, but also makes the encrypted ROI more complete and secure.

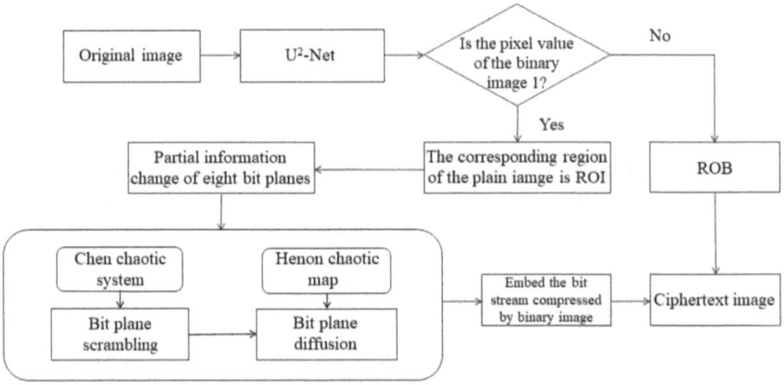

Fig. 2. The encryption framework diagram of ROI in medical images

Encryption Steps:

(1) Input gray image I.
(2) Process I using U^2-Net to obtain the binary image B, where a pixel value of 0 corresponds to the ROB region in I, and a pixel value of 1 corresponds to the ROI region in I. And obtain the position information of ROI according to the binary image $(X, Y) = \left[(x_1, y_1), (x_2, y_2), \ldots, (x_n, y_n)\right]$.
(3) Act on the position information (X, Y) to I to get the image Img (all pixel values within the Img are zero except ROI).

(4) Decompose Img into eight bit planes $PS\{i\}$, $(i = 1, 2, \ldots, 8)$. Since the fourth bit plane is similar to the fifth bit plane in terms of light and darkness, part information of PS{4} and PS{5} is exchanged.
(5) Use PS{1}, PS{2} and PS{3} to control the contents of PS{4}, PS{5} and $PS\{7\}$ to get $PS1\{i\}$, $(i = 1, 2, \ldots, 8)$ (This is because the information distribution in the first, second and third planes is relatively uniform, so they are used to control the fourth, fifth and seventh planes with high information correlation in the bits. At the same time, there is too much information in the sixth bit plane and too little information in the eighth bit plane, so no control operations are performed on these two bit planes).
(6) Perform plane pixel scrambling on $PS1\{i\}$, $(i = 1, 2, \ldots, 8)$ using Chen chaotic system. According to Eq. (2), three sets of keys K1, K2 and K3 are generated, and the bit plane scrambling is performed on $PS1\{4\}$, PS{5}, PS{6} and $PS\{7\}$ to obtain $PS2\{i\}$, $(i = 1, 2, \ldots, 8)$. (Partial scrambling of bit planes can reduce the running time and improve the encryption efficiency).
(7) Use Henon chaotic map to generate the key matrix P with the same size as I, and decompose P into eight bit planes. After the decomposition, $PS3\{i\}$, $(i = 1, 2, \ldots, 8)$ is obtained by XOR operation corresponding to the eight bit planes of $PS2$ one by one.
(8) $PS3\{i\}$, $(i = 1, 2, \ldots, 8)$ were bitwise recovered into image E and combined with the unaltered ROB (the ROB region exhibits unstructured noise characteristics, and its information entropy value is significantly lower than the effective information threshold of medical images. Despite the presence of macroscopic contour feature residues in the ROI region, its texture contrast and correlation do not reach the minimum threshold for recognizable anatomical structures, which is insufficient to support the accurate identification of anatomical structures) to obtain image E1.
(9) The binary image B is compressed using run-length encoding and converted into a bit stream K. The key K1 of Chen's chaotic system is used to control the position of the bit stream embedded in the image. The final image EI is obtained.

3.2 Decryption Scheme

The encryption scheme compresses the binary image using run-length encoding and embeds the image through key control. It weakens the possibility of image cracking, saves the space needed to store the binary image separately during transmission, and more importantly, enables lossless decryption of medical images. Figure 3 shows the structure of medical image ROI lossless decryption.

Decryption Steps:

(1) Extract the bitstream K from the ciphertext image EI using key X1 to get the image E1.
(2) Decompress the bitstream K into a binary image B using the inverse operation of run-length encoding.
(3) The image E is obtained from the correspondence between the binary image B and the ROI within E1 and decomposed into the eight bit planes $PS3\{i\}$, $(i = 1, 2, \ldots, 8)$.
(4) The key matrix P is XORed to the eight matrices of $PS3\{i\}$, $(i = 1, 2, \ldots, 8)$, one by one after bit plane decomposition to obtain $PS2\{i\}$, $(i = 1, 2, \ldots, 8)$.

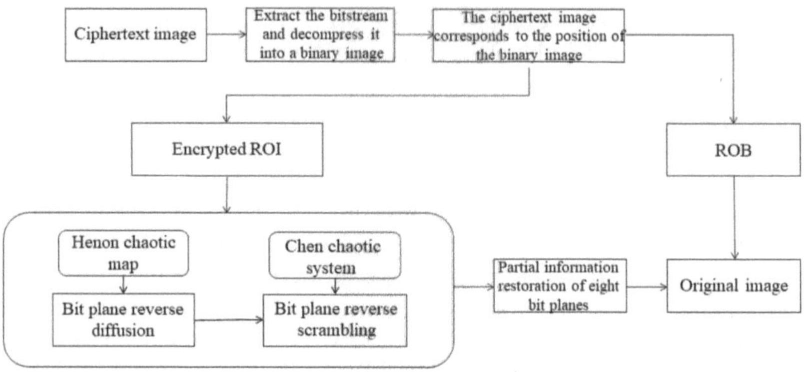

Fig. 3. The decryption framework diagram of ROI in medical images

(5) Using the keys K1, K2 and K3, perform inverse scrambling operations in the bit plane on $PS2\{4\}$, $PS2\{5\}$, $PS2\{6\}$ and $PS2\{7\}$ to obtain $PS1\{i\}$, $(i = 1, 2, \ldots, 8)$.
(6) Use PS{1}, PS{2} and PS{3} to reverse the control of $PS1\{4\}$, $PS1\{5\}$ and $PS1\{7\}$, with $PS1\{6\}$ and $PS1\{8\}$ remaining unchanged, to get $PS\{i\}$, $(i = 1, 2, \ldots, 8)$. And exchange part of the information of $PS\{4\}$ and $PS\{5\}$.
(7) Recover the image Img from $PS\{i\}$, $(i = 1, 2, \ldots, 8)$ by bit, and reconstruct the image DI with the unchanged ROB to complete the decryption.

Note: The above encryption and decryption scheme is also applicable to the color image, which is converted into a gray image for U^2-Net processing, to determine the ROI and ROB regions of the color image, and then encrypt and decrypt the color image by channel in accordance with the above encryption and decryption scheme.

4 Comparison and Analysis of Experimental Results

This experiment is simulated under MATLAB 2016a and Python to evaluate the proposed encryption method. All data and encryption tests were obtained from real-world operations on a computer with AMD R7-5800H 3.20 GHz CPU and 16 GB RAM and Windows 11 environment. In order to test the accuracy of ROI extraction and the effectiveness of image encryption and decryption, 342 images of four different head diseases were tested in the experiment. The data source is mentioned in the reference [29].

4.1 Simulation Results

Four of 342 images were randomly selected in Fig. 4. The sizes of these four images are all 512×512. In addition, the control parameters of the Chen chaotic system during the simulation are $a = 35$, $b = 3$, and $c = 28$, and the three initial values are $x_0 = 0$, $y_0 = 1$, and $z_0 = 0$. The control parameters of the Henon chaotic mapping are $p = 1.4$, $q = 0.3$, and the initial values are $s_0 = 0.2$ and $t_0 = 0.2$.

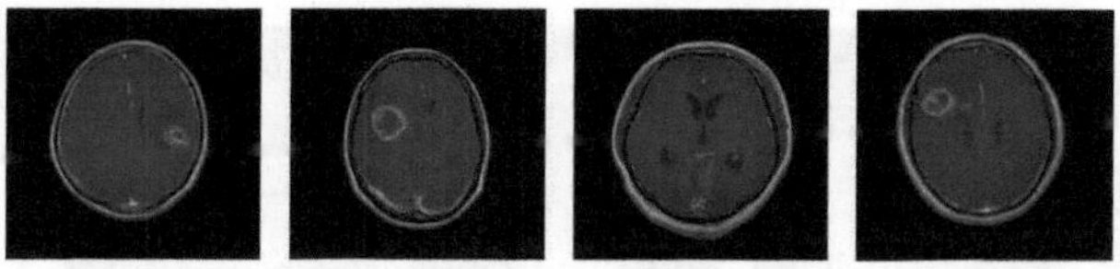

a) Org 1 b) Org 2 c) Org 3 d) Org 4

Fig. 4. Part of original images

Figure 5 shows the binary images extracted by U^2-Net, LOG-OLOH [30] and MCS [31] respectively. From the figure, it can be noticed that compared to the other two algorithms, the edges of the binary image extracted by U^2-Net are smoother and the extracted results are more accurate. This is due to U^2-Net learns and improves continuously during the training process. Therefore, it is reasonable to apply U^2-Net to medical image ROI extraction.

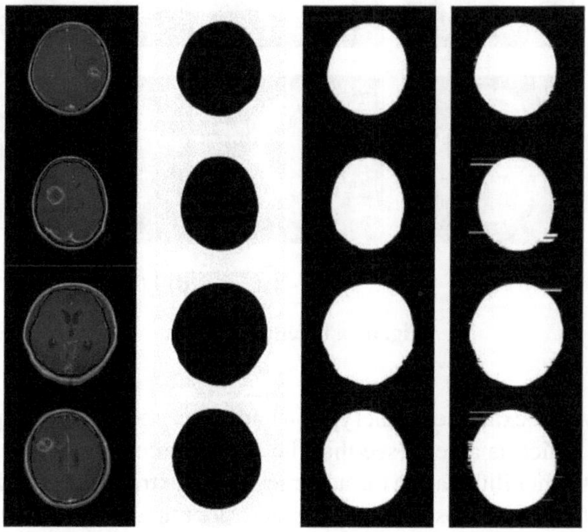

a) Original b) U^2-Net c)LOG-OLOH d) MCS

Fig. 5. Binary images

Figure 6 shows the results of comparing the ciphertext images of the four images in the example with LOG-OLOH, MCS and LSB-LWT [32]. From Fig. 6, the cipher image obtained by LOG-OLOH is relatively accurate, but the proposed algorithm encrypts the ROI more completely, and the ROI encrypted by methods MCS and LSB-LWT has irregular edges. From the effect of ciphertext image, the image encryption is better after accurate and complete extraction of ROI by U^2-Net.

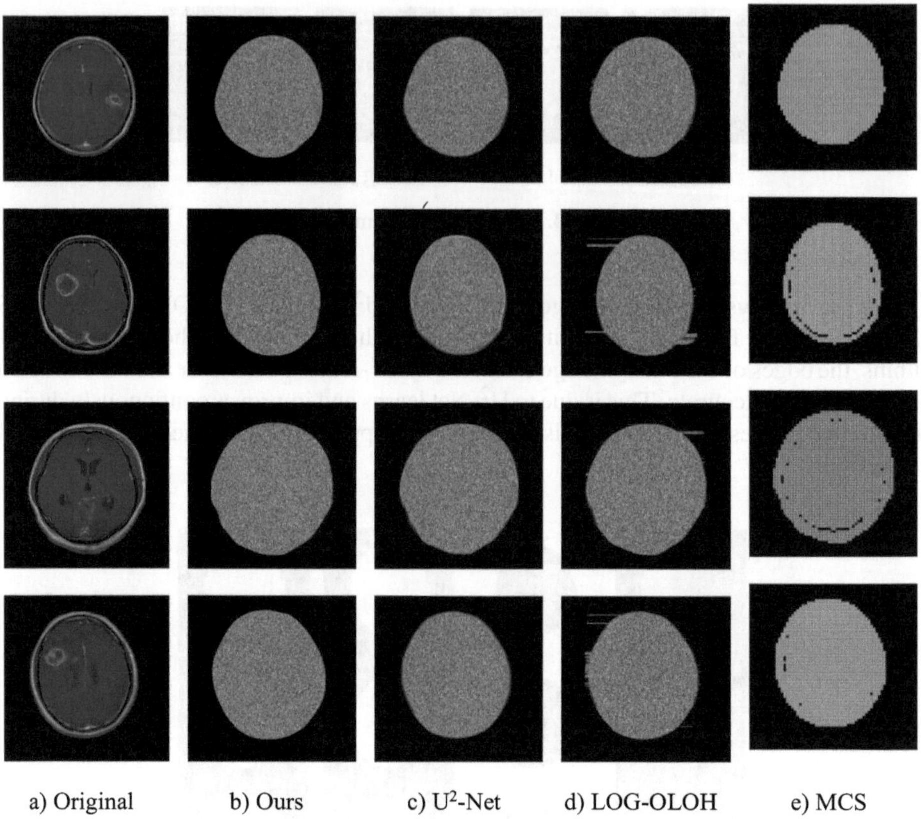

Fig. 6. Ciphertext images

Figure 7 shows an example of encryption and decryption of the proposed algorithm. From the figure, which is able to see that the algorithm encrypts only the ROIs of the medical images, further illustrating the accuracy of the extraction. In addition, the images in the first and third columns of the figure are identical, explaining that decryption is lossless. This is because the binary image is compressed and embedded into the encrypted image after key control. Extracting the secret information from the ciphertext image can accurately extract the encrypted ROI, so as to achieve lossless decryption. The quantitative analysis of this algorithm will be proceeded next to further verify it is feasible.

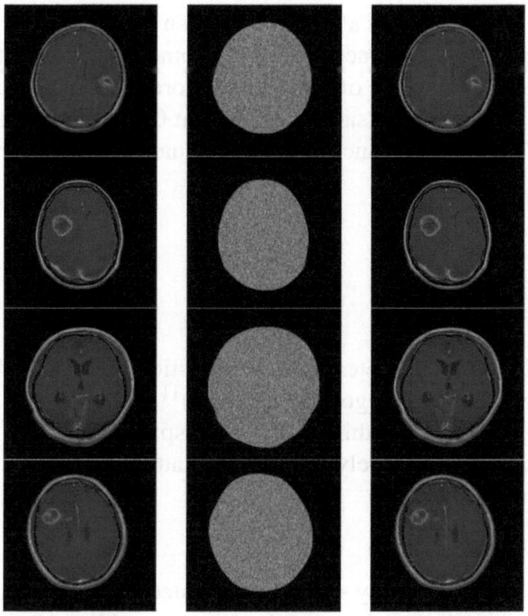

a) Original b) Encryption c) Decryption

Fig. 7. Encryption and decryption

4.2 Running Time

The encryption efficiency of the method is relatively low due to the partial time spent in ROI extraction. The time taken by the proposed algorithm to complete ROI region extraction and the whole encryption process is shown in detail in Table 1 with the running time of LOG-OLOH, MCS and LSB-LWT and the average running time of the four images is calculated.

Table 1. Running Time(s)

Image Name	Ours	LOG-OLOH	MCS	LSB-LWT
Org 1	0.575661	0.190824	0.197646	0.082550
Org 2	0.557243	0.193610	0.219822	0.079766
Org 3	0.586801	0.206373	0.202132	0.081067
Org 4	0.573251	0.199812	0.198990	0.080914
Mean Time	0.573240	0.197650	0.204650	0.081074

As Table 1, this algorithm runs a little longer than the other algorithms. This means that the accurate and complete extraction of ROIs comes with the cost of time consumption, with an average running time of 0.375590 s more than that of LOG-OLOH. From the current analysis it seems that "sacrificing" about 0.38 s improves the completeness of ROI extraction. Other performances will be evaluated to determine if the extra time is worth it.

4.3 Key Analysis

1. Key space analysis

This algorithm uses 11 parameters in the encryption process, and they are independent. So key space value of this algorithm is $(10^{15})^{11} = 10^{165} \approx 2^{548}$ [33]. According to the capability of computers at this stage, a key space size of 2^{128}indicates that the ciphertext image is able to effectively resist violent attacks. Thus, this algorithm is able to endure the exhaustive attack.

2. Key sensitivity analysis

Taking Org 1 as an example, key sensitivity is tested using the four parameters x_0, y_0, a and b. In the simulation, the four keys are decrypted by adding 10^{-15} to each of them to demonstrate that the subtle movements of the keys can significantly affect the decryption results. These decrypted images obtained using "tampering" keys are displayed in Fig. 8.

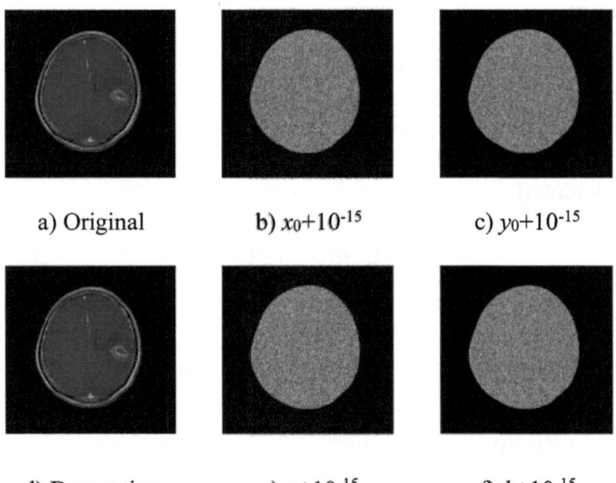

Fig. 8. Test of key sensitivity

As Fig. 8, even the slightest change in any key is able to enormously influence decryption result, i.e., it cannot be recovered as a plaintext image. This indicates that the key is sensitive.

4.4 Statistical Attack Analysis

In this subsection, this algorithm feasibility is analysed through histogram analysis, Shannon entropy and adjacent pixels correlation.

1. Histogram analysis

The histograms of three states of four images in example are shown in Fig. 9. The first two columns of the histogram have significant difference in Fig. 9, where original image pixel distribution in the first column shows a high and low state, and the pixels of the encrypted image in the second column are almost uniformly distributed. Furthermore, the histogram of the plaintext and decrypted image are identical. The analysis of this histogram results shows that this method is able to effectively avoid statistical hazards.

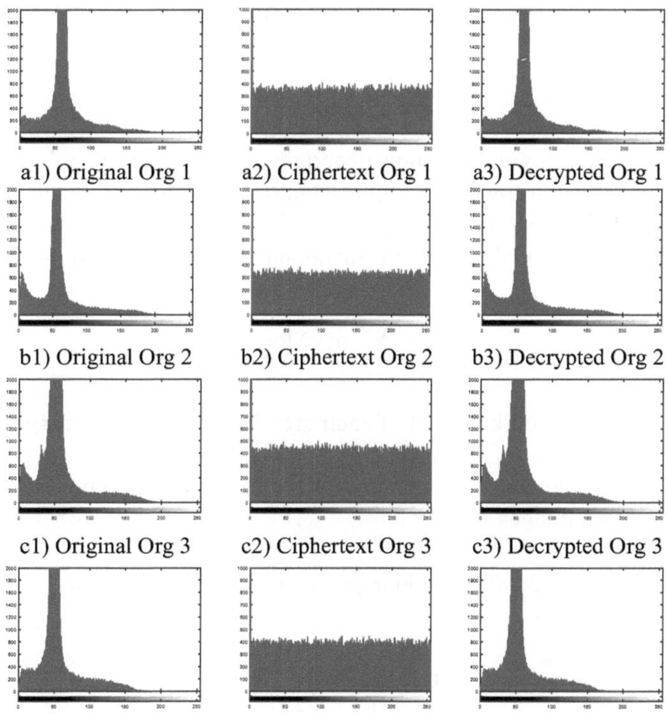

a1) Original Org 1 a2) Ciphertext Org 1 a3) Decrypted Org 1

b1) Original Org 2 b2) Ciphertext Org 2 b3) Decrypted Org 2

c1) Original Org 3 c2) Ciphertext Org 3 c3) Decrypted Org 3

d1) Original Org 4 d2) Ciphertext Org 4 d3) Decrypted Org 4

Fig. 9. Histogram of original, ciphertext and decrypted images

The variance is calculated by Eq. (3) [34, 35] to quantitatively analyse histograms distribution.

$$\mathrm{var}(X) = \frac{1}{N^2}\sum_{i=0}^{N-1}\sum_{j=0}^{N-1}\frac{1}{2}(x_i - x_j)^2 \tag{3}$$

where $X = \{x_1, x_2, \ldots, x_N\}$ is the histogram vector, $N = 256$. x_i and x_j are pixels number with grey value i and j respectively. Table 2 shows the histogram variance of the original and cipher images. By the above proofs this method is able to resist the histogram attack.

Table 2. Histogram Variance of the Original and Ciphertext Image

Image Name	Original Image	Ciphertext Image	Decrypted image
Org 1	1.0150e+06	2.8691e+04	1.0150e+06
Org 2	8.0665e+05	2.0981e+04	8.0665e+05
Org 3	8.8801e+05	5.4480e+04	8.8801e+05
Org 4	9.9621e+05	3.9517e+04	9.9621e+05

2. Shannon entropy

Shannon entropy helps quantify how uncertain or random an image's pixel values are [36]. Higher Shannon entropy near 8 implies random pixel distribution in cipher images. The likelihood of the encrypted image being decrypted following an entropy-based assault is fairly minimal. Calculation of Shannon entropy [37] as Eq. (4).

$$H(s) = \sum_{i=0}^{M-1} p(s_i)\log_2\frac{1}{p(s_i)} \tag{4}$$

where $p(s_i)$ represents the likelihood of each grey level within the encrypted image, $M = 256$.

Table 3 displays the Shannon entropy for the example, encrypted images, and the mean entropy of 342 images.

Table 3. Shannon Entropy of Original and Ciphertext Images

Image Name	Original Image	Ciphertext Image
Org 1	5.8705	7.9989
Org 2	6.0335	7.9989
Org 3	6.3562	7.9986
Org 4	6.0924	7.9983
Mean Value	–	7.9988

Table 3 shows the Shannon entropy values of four encrypted images are near 8. The mean entropy value for 342 ciphertext images in the experiment is 7.9988, indicating the designed algorithm is able to withstand entropy-based attacks. This result of entropy is because bit planes decomposition is done first in the encryption process and then some of the bit planes are encrypted using the chaotic system. Table 4 lists the Shannon entropy of the proposed algorithm against several different encryption algorithms.

Table 4. Comparison of Shannon Entropy with other encryption algorithms

Image Name	Ours	LoG-OLHL	SMLS [38]	MCS	LSB-LWT
Org 1	7.9989	7.9978	6.8950	7.9977	6.1294
Org 2	7.9989	7.9980	6.6992	7.9978	6.2927
Org 3	7.9986	7.9984	7.0008	7.9984	6.4348
Org 4	7.9983	7.9980	7.0213	7.9983	6.4055
Mean Value	7.9987	7.9981	6.9041	7.9981	6.3156

Table 4 shows Shannon and average entropy values for five encryption algorithms. Comparison indicates the average entropy of this chapter's algorithm surpasses the other four, demonstrating the proposed scheme's resilience to entropy attacks.

3. Adjacent pixels correlation

The correlation coefficient of neighboring pixels measures their relationship. Reducing this strong correlation effectively protects against statistical attacks. Ideally, a ciphertext image's correlation coefficient should be near zero. Two-dimensional correlation analysis, as detailed in Eq. (5) [39], measures this adjacency correlation.

$$r_{xy} = \frac{\mathrm{cov}(x, y)}{\sqrt{D(x)D(y)}} \tag{5}$$

Among them:

$$\mathrm{cov}(x, y) = \frac{1}{N}\sum_{i=1}^{N}(x_i - E(x))(y_i - E(y)),\ (i = 1, 2, \ldots, N) \tag{6}$$

$$E(x) = \frac{1}{N}\sum_{i=1}^{N}x_i,\ (i = 1, 2, \ldots, N) \tag{7}$$

$$D(x) = \frac{1}{N}\sum_{i=1}^{N}(x_i - E(x))^2,\ (i = 1, 2, \ldots, N) \tag{8}$$

where x and y represent two neighboring pixels, and r_{xy} stands for their correlation coefficient.

This section analyses adjacent pixels correlation using Org 1. $N = 100000$ pairs of adjacent pixels are extracted horizontally, vertically and diagonally within their ROI and their r_{xy} is calculated. Figure 10 shows the correlation of N pairs of pixels within the ROI of the plaintext and ciphertext image.

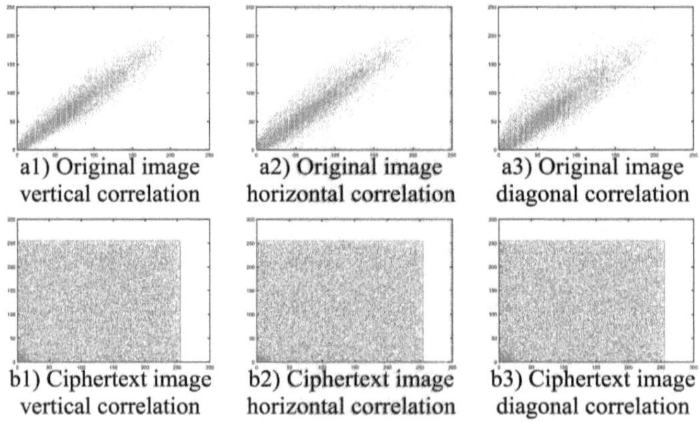

Fig. 10. Histograms in vertical, horizontal and diagonal directions of original and ciphertext images

From Fig. 10, it can be noticed that the correlation patterns of original and encrypted images clearly inconsistent even in the same direction. Before encryption, adjacent pixels in the horizontal, vertical and diagonal directions are strongly correlated. Following encryption, correlations in all three directions within the ROI become weak. Table 5 presents the r_{xy} values for the four images in all three directions.

Table 5. Correlation Coefficient Before and After Encryption

Image Name	State	Vertical	Horizontal	Diagonal
Org 1	Original Image	0.9393	0.9110	0.8546
	Ciphertext Image	−0.0112	−0.0057	−0.0016
Org 2	Original Image	0.9337	0.9218	0.8536
	Ciphertext Image	−0.0043	0.0094	−0.0085
Org 3	Original Image	0.9339	0.9457	0.8876
	Ciphertext Image	−0.0097	0.0054	−0.0079
Org 4	Original Image	0.9323	0.9171	0.8539
	Ciphertext Image	−0.0046	0.0034	0.0068

An absolute correlation coefficient of zero between adjacent pixels indicates optimal image encryption. Table 5 shows the original image ROI's correlation coefficient is near 1, while the encrypted ROI's is minimal, implying that the two adjacent pixels within the ciphertext ROI are extremely uncorrelated. Table 6 compares the proposed algorithm's correlation coefficients with LOG-OLOH, SMLS, MCS and LSB-LWT with respect to image Org 1.

Table 6. Results of Correlation Coefficient Comparison

Algorithm	Vertical	Horizontal	Diagonal
Ours	−0.0112	−0.0057	−0.0016
Algorithm [30]	−0.0071	−0.0023	0.0019
Algorithm [39]	0.0504	0.0646	0.0252
Algorithm [31]	−0.0037	−0.0068	−0.0046
Algorithm [32]	0.0189	0.2259	0.1364

Table 6 indicates that the proposed algorithm, along with LOG-OLOH and MCS, exhibits weak correlation. The data in the table demonstrates the proposed algorithm's superior effectiveness in eliminating pixel correlation.

4.5 Differential Attack Analysis

The Number of Pixels Change Rate (NPCR) measures the proportion of differing pixel values at identical positions between two images of equal size relative to the total pixel count. Unified Average Change Intensity (UACI) typically quantifies the mean intensity difference between two images, pre- encryption and post-encryption.

The ideal UACI value is 33.4635%, and for NPCR, it is 99.6094%. Equation (10) and Eq. (11) present the formulas for NPCR and UACI [40, 41].

$$\mathrm{NPCR} = \frac{\sum_{i=1}^{M} D(i)}{M} \times 100\%,\ (i = 1, 2, \ldots, M) \tag{9}$$

$$\mathrm{UACI} = \frac{1}{M}\sum_{i=1}^{M}\frac{|C_1(i) - C_2(i)|}{255} \times 100\%,\ (i = 1, 2, \ldots, M) \tag{10}$$

$$D(i) = \begin{cases} 0, C_1(i) = C_2(i) \\ 1, C_1(i) \neq C_2(i) \end{cases},\ (i = 1, 2, \ldots, M) \tag{11}$$

where $C_1(i)$ and C_2 (i) represent two ROI sequences, one before and one after image encryption. M indicates the sequence length.

Table 7 presents the NPCR and UACI values for the cipher image's ROI sequence.

Table 7. NPCR (%) and UACI (%)

Image Name	NPCR	UACI
Org 1	0.996	0.3271
Org 2	0.9959	0.3327
Org 3	0.9961	0.3568
Org 4	0.9961	0.3433
Mean Value	0.9961	0.3362

As can be seen from Table 7, the NPCR and UACI of all four images are close to the ideal values. The two averages of 342 images in the dataset approach these ideal values of 99.61% and 33.62%, respectively, demonstrating the ROI encryption algorithm for medical images in this chapter can withstand differential attacks effectively. Tables 8 and 9 display the comparative NPCR and UACI results.

Table 8. Comparison of NPCR (%) with Other Algorithms

Image Name	Ours	LOG-OLOH	MCS	LSB-LWT
Org 1	0.996	0.9961	0.9954	0.991
Org 2	0.9959	0.9962	0.9957	0.9935
Org 3	0.9961	0.9957	0.9961	0.9953
Org 4	0.9961	0.9958	0.9961	0.9952
Mean Value	0.99602	0.99595	0.99582	0.99375

Table 9. Comparison of UACI (%) with Other Algorithms

Image Name	Ours	LOG-OLOH	MCS	LSB-LWT
Org 1	0.3271	0.3222	0.3214	0.3203
Org 2	0.3327	0.3382	0.3402	0.3028
Org 3	0.3568	0.3412	0.3397	0.313
Org 4	0.3433	0.3387	0.339	0.3018
Mean Value	0.33997	0.33508	0.33508	0.30948

Since SMLS encrypts the ROB region, only the proposed algorithm is evaluated against LOG-OLOH, MCS and LSB-LWT. NPCR and UACI scores are computed for each of four images and their averages. The comparison reveals that the two indexes of this algorithm are closer to optimal values, indicating this algorithm has better encryption performance.

4.6 Structure Similarity Analysis

Structure similarity index measure (SSIM) serves as a key criterion for discriminating image similarity. Its calculation method [42] is shown in Eq. (13).

$$SSIM\,(x, y) = \left[\frac{(2\mu_x\mu_y + c_1)(2\sigma_{xy} + c_2)}{(\mu_x^2 + \mu_y^2 + c_1)(\sigma_x^2 + \sigma_y^2 + c_2)} \right] \tag{12}$$

$$c_1 = k_1^2 L^2 \tag{13}$$

$$c_2 = k_2^2 L^2 \tag{14}$$

where x and y represent ROI sequences from the original and encrypted images respectively. μ_x and μ_y are their average values, respectively. Sequences variances x and y are denoted by σ_x^2 and σ_y^2, respectively. σ_{xy} denotes the covariance between x and y. L refers to the pixel value range. k_1 and k_2 take values 0.01 and 0.03.

Table 10 lists the SSIM values for the four images.

Table 10. SSIM

Image Name	SSIM
Org 1	0.0137
Org 2	0.0124
Org 3	0.0108
Org 4	0.012
Mean Value	0.012

Table 10 reveals low SSIM values for all four example images. Across the 342 experimental images, the average SSIM was 0.0120, showing extreme dissimilarity between the pre- and post-encryption ROI sequences. These findings further substantiate the viability of the proposed encryption method. Table 11 presents SSIM value comparisons between the proposed algorithm and others.

Table 11. Comparison of SSIM

Image Name	Ours	LoG-OLHL	SMLS	MCS	LSB-LWT
Org 1	0.0137	0.0138	0.0031	0.5505	0.0867
Org 2	0.0124	0.0120	0.0032	0.5575	0.1112
Org 3	0.0108	0.0113	0.0032	0.4616	0.1169
Org 4	0.0120	0.0130	0.0032	0.4974	0.0968
Mean Value	0.0122	0.0125	0.0032	0.5168	0.1029

Table 11 shows the SSIM values and average values calculated for each of the four images using the five algorithms. Among them, SMLS has the smallest SSIM value, which is caused by its method of encrypting the ROB and subsequently scrambling the encrypted ROB and ROI blocks. Furthermore, the SSIM of the encryption algorithm presented here is lower than that of comparable algorithms like LoG-OLHL, MCS, and LSB-LWT. Experiments have proved that the encrypted image produced by this algorithm has relatively less similarity with the original image.

4.7 Lossless Decryption Analysis

The encryption scheme incorporates run-length encoding, and secret information is embedded into encrypted image using a key to ensure lossless decryption as well as security of the image. Following figure shows the decrypted image obtained without embedding the secret information. Figure 11 illustrates the comparison findings.

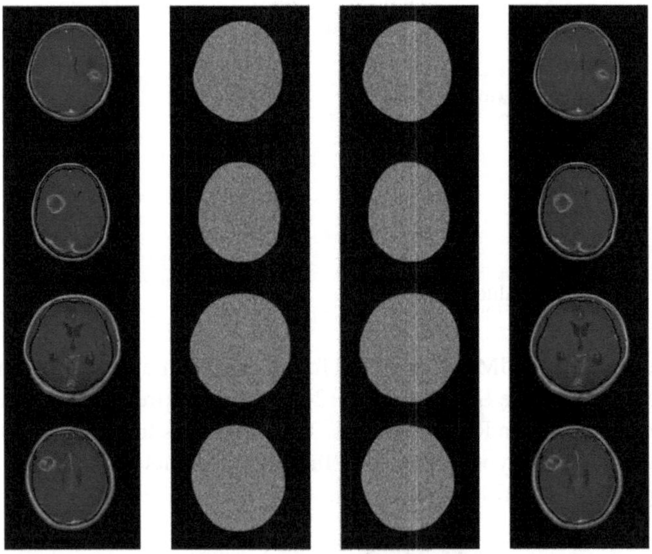

a) Original image | b) Ciphertext Image | c) Decrypted image without embedded secret information | d) True decrypted image

Fig. 11. Comparison of decryption results

Figure 11 clearly shows that the image resulting from omitting ROI extraction, based on the secret information during decryption, bears no resemblance to the original image. This phenomenon is caused by the inconsistency between the ROI region obtained during decryption and the encrypted ROI region, which makes the decrypted sequence inconsistent with the original sequence. To address this issue and enable exact medical image recovery, binary images are compressed using run-length encoding and embedded into the encrypted image under key control. This not only ensures medical images lossless decryption, but also compressing the information optimizes space utilization for storage and transmission and saves embedding time. In order to quantitatively analyse that the proposed algorithm is lossless decryption, it is proved by experimentally differing the original from decrypted image to obtain a zero matrix. Furthermore, the first and last columns of the corresponding images show no substantial difference in Fig. 11 also proves that the decryption is lossless.

5 Conclusion

In this paper, a highly secure reversible selective encryption algorithm for medical images based on bit plane and controlled compression is proposed. Firstly, a binary image is obtained by applying U^2-Net to the medical image. Second, the ROI of medical image is extracted by using the positional relationship between the binary image and the ROI in the original image, and then the medical image ROI bit planes are decomposed and encrypted to ensure the security of ROI. Finally, the binary image is compressed using run-length encoding and embedded into the encrypted image using the key to achieve lossless decryption. According to the simulation experiments, although the encryption algorithm spends some time in extracting the ROIs, it obtains more complete, more accurate and smoother ROIs, and the encrypted images are more secure. The design of this algorithm balances some of the running time by encrypting some of the bit planes and compressing the secret information. The algorithm is proved to be feasible based on the analysis.

For protecting medical images information more securely, based on the existing research, future research work is explored in the following points:

(1) U^2-Net network training time is too long. Further research on the application of machine learning related methods in image processing, so as to shorten the running time of ROI extraction is the next problem to be solved.
(2) Taking chaotic systems as a starting point, designing new chaotic systems with good chaotic properties through new ideas is the focus of the next step.
(3) Cutting from multi-image encryption, designing multi-image medical image ROI encryption algorithms is also a worthwhile research issue.

Acknowledgments. This work is supported by the Shandong Provincial Natural Science Foundation (CN) (ZR2022LLZ003) and the National Natural Science Foundation of China (12471422).

Disclosure of Interests. The authors declare that they have no known competing financial interests or personal relationships that could have appeared to influence the work reported in this paper.

References

1. Wu, J., Liao, X., Yang, B.: Color image encryption based on chaotic systems and elliptic curve elgamal scheme. Signal Process. **141**, 109–124 (2017)

2. Zhou, J., Li, J., Di, X.: A novel lossless medical image encryption scheme based on game theory with optimized ROI parameters and hidden ROI position. IEEE Access **8**, 122210–122228 (2020)
3. Abd EL-Latif, A.A., Abd-El-Atty, B., Abou-Nassar, E.M., et al.: Controlled alternate quantum walks based privacy preserving healthcare images in internet of things. Opt. Laser Technol. **124**, 105942 (2020)
4. Itti, L., Koch, C., Niebur, E.: A model of saliency-based visual attention for rapid scene analysis. IEEE Trans. **20**(11), 1254–1259 (1998)
5. Cheng, M.M., Mitra, N.J., Huang, X., et al.: Global contrast based salient region detection. Comput. Vis. Pattern Recogn. **37**(3), 569–582 (2015)
6. Pandey, D., Yin, X., Wang, H., et al.: Automatic and fast segmentation of breast region-of-interest (ROI) and density in MRIs. Heliyon **4**(12), 111–130 (2018)
7. Gaidel, A.: Method of automatic ROI selection on lung CT images. Procedia Eng. **201**, 258–264 (2017)
8. Vig, E., Dorr, M., Cox, D.: Large-scale optimization of hierarchical features for saliency prediction in natural images. In: IEEE Conference on Computer Vision and Pattern Recognition, Columbus, pp. 2798–2805 (2014)
9. Liu, N., Han, J., Zhang, D., et al.: Predicting eye fixations using convolutional neural networks. In: Proceedings of IEEE Conference on Computer Vision and Pattern Recognition, Boston, pp. 362–370 (2015)
10. Kruthiventi, S.S.S., Ayush, K., Babu, R.V.: DeepFix: a fully convolutional neural network for predicting human eye fixations. IEEE Trans. Image Process. **26**(9), 4446–4456 (2017)
11. Pan, J., Sayrol, E., Giro-i-Nieto, X., et al.: Shallow and deep convolutional networks for saliency prediction. In: Proceedings of IEEE Conference on Computer Vision and Pattern Recognition, pp. 598–606 (2016)
12. Bhatnagar, G., Wu, Q.M.J.: Selective image encryption based on pixels of interest and singular value decomposition. Digit. Signal Process. **22**(4), 648–663 (2012)
13. Wu, Y., Noonan, J.P., Agaian, S.: A Wheel-switch chaotic system for image encryption. In: Proceedings 2011 International Conference on System Science and Engineering, Macau, pp. 23–27 (2011)
14. Chen, J., Zhu, Z., Fu, C., et al.: A fast image encryption scheme with a novel pixel swapping-based confusion approach. Nonlinear Dyn. **77**(4), 1191–1207 (2014)
15. Wang, X., Liu, L., Zhang, Y.: A novel chaotic block image encryption algorithm based on dynamic random growth technique. Opt. Lasers Eng. **66**, 10–18 (2015)
16. Xu, L., Gou, X., Li, Z., et al.: A novel chaotic image encryption algorithm using block scrambling and dynamic index based diffusion. Opt. Lasers Eng. **91**, 41–52 (2017)
17. Mozaffari, S.: Parallel image encryption with bitplane decomposition and genetic algorithm. Multimed. Tools Appl. **77**, 25799–25819 (2018)
18. Som, S., Mitra, A., Palit, S., et al.: A selective bitplane image encryption scheme using chaotic maps. Multimed. Tools Appl. **78**, 10373–10400 (2019)
19. Yin, Z., Peng, Y., Xiang, Y.: Reversible data hiding in encrypted images based on pixel prediction and bit-plane compression. IEEE Trans. Dependable Secure Comput. **19**(2), 992–1002 (2020)
20. Ronneberger, O., Fischer, P., Brox, T.: U-Net: convolutional networks for biomedical image segmentation. In: Medical Image Computing and Computer-Assisted Intervention–MICCAI 2015: 18th International Conference, Germany, pp. 234–241 (2015)
21. Qin, X., Zhang, Z., Huang, C., et al.: U2-net: going deeper with nested u-structure for salient object detection. Pattern Recogn. **106**, 107404 (2020)
22. Xie, S., Tu, Z.: Holistically-nested edge detection. In: 2015 IEEE International Conference on Computer Vision (ICCV), Santiago, pp. 1395–1403 (2015)

23. Xu, J., Zhao, B., Wu, Z.: Research on color image encryption algorithm based on bit-plane and chen chaotic system. Entropy **24**(2), 186 (2022)
24. Danial, R., Mahdi, Y.: Color image encryption using hyper chaos chen. Int. J. Comput. Appl. **110**, 9–12 (2015)
25. Al-Hazaimeh, O., Al-Jamal, M., Bawaneh, M., et al.: A new image encryption scheme using dual chaotic map synchronization. Int. Arab J. Inf. Technol. **18**(1), 95–102 (2021)
26. Henon, M.: A two-dimensional mapping with strange attractor. Commun. Math. Phys. **50**(1), 376–392 (1976)
27. Hu, Y., Wu, H., Zhou, L.: Color image encryption base on a 2D hyperchaotic enhanced henon map and cross diffusion. Alex. Eng. J. **73**, 385–402 (2023)
28. Hou, A.L., Yuan, F., Ying, G.: QR code image detection using run-length coding. In: Proceedings of 2011 International Conference on Computer Science and Network Technology, Harbin, pp. 2130–2134 (2011)
29. Andrew, N.: Stanford Machine Learning Group. Stanford: Stanford ML Group, 2021, 26 October 2022. https://stanfordmlgroup.github.io/
30. Zhang, L., Song, X., El-Latif, A.A.A., et al.: Reversibly selective encryption for medical images based on coupled chaotic maps and steganography. Complex Intell. Syst. 1–27 (2023)
31. Kiran, P., Parameshachari, B.D.: Resource optimized selective image encryption of medical images using multiple chaotic systems. Microprocess. Microsyst. **91**, 1–21 (2022)
32. Liu, X., Chen, H.Q., Xia, W.G.: Selective medical image encryption based on LSB and lifting wavelet transform. Intell. Comput. Appl. **13**(2), 139–144 (2023)
33. Zhu, S., Zhu, C.: Plaintext-related image encryption algorithm based on block structure and five-dimensional chaotic map. IEEE Access **7**, 147106–147118 (2019)
34. Khan, J.S., Boulila, W., Ahmad, J., et al.: DNA and plaintext dependent chaotic visual selective image encryption. IEEE Access **8**, 159732–159744 (2020)
35. Lu, Q., Zhu, C., Deng, X.: An efficient image encryption scheme based on the LSS chaotic map and single S-box. IEEE Access **8**, 25664–25678 (2020)
36. Shannon, C.E.: Communication theory of secrecy systems. Bell Syst. Tech. J. **28**(4), 656–715 (1949)
37. Madhusudhan, K.N., Sakthivel, P.: A secure medical image transmission algorithm based on binary bits and Arnold map. J. Ambient. Intell. Humaniz. Comput. **12**, 5413–5420 (2021)
38. Sankaradass, V., Murali, P., Tholkapiyan, M.: Region of interest (ROI) based image encryption with sine map and Lorenz system. In: Proceedings of the International Conference on ISMAC in Computational Vision and Bio-Engineering 2018 (ISMAC-CVB), pp. 493–502. Springer, Cham (2019)
39. Teng, L., Wang, X., Meng, J.: A chaotic color image encryption using integrated bit-level permutation. Multimed. Tools Appl. **77**, 6883–6896 (2018)
40. Ramasamy, P., Ranganathan, V., Kadry, S., et al.: An image encryption scheme based on block scrambling, modified zigzag transformation and key generation using enhanced logistic-tent map. Entropy **21**(7), 656 (2019)
41. Ahmad, J., Larijani, H., Emmanuel, R., et al.: Secure occupancy monitoring system for IoT using lightweight intertwining logistic map. In: 2018 10th Computer Science and Electronic Engineering (CEEC), Colchester, pp. 208–213 (2018)
42. Abd-El-Atty, B., Iliyasu, A.M., Alaskar, H., et al.: A Robust quasi-quantum walks-based steganography protocol for secure transmission of images on cloud-based e-healthcare platforms. Sensors **20**(11), 3108 (2020)

Edge Intelligence

Accelerating Large-Scale GNN by Combining k-Hop Neighbors Based Feature Caching and Hierarchical GPU-Centric Data Access

Wenbin Jiang[1,2(✉)], Fanxing Pan[1], Rui Wu[1], Xinhai Shen[1], Dongao He[1], and Hai Jin[1]

[1] National Engineering Research Center for Big Data Technology and System, Services Computing Technology and System Lab, Cluster and Grid Computing Lab, School of Computer Science and Technology, Huazhong University of Science and Technology, Wuhan 430074, Hubei, China
wenbinjiang@hust.edu.cn

[2] Zhejiang Laboratory, Hangzhou 311121, Zhejiang, China

Abstract. Graph neural networks (GNNs) have been widely applied in graph-based learning tasks. However, the feature collection stage is increasingly becoming the bottleneck of the existing GNN systems when training large-scale graphs. To solve this problem, this paper proposes FCGraph, a novel efficient GNN training strategy by combining a feature cache policy of multi-hop (k-hop) neighbors and a hierarchical GPU-centric data access method. First, a k-hop neighbors based cache policy is proposed by exploring vertex data access characteristics in k-layer GNN models, which focuses on reducing CPU-GPU data transfer overhead. Second, based on the cache policy, node features are partitioned according to node access frequency, and a novel GPU-centric data access method is presented for feature collection. Further, FCGraph is scaled to multi-GPU systems equipped with NVLink, in which the memory access hierarchies are explored. The evaluation of some representative datasets shows that FCGraph can improve end-to-end training performance over the advanced GNN training systems of DGL and PyTorch-Direct by 3.72× and 1.46× on average, respectively.

Keywords: Deep Learning · Graph Neural Network · Feature Cache · GPU

1 Introduction

Graph neural networks (GNNs) have achieved great success on various graph-based learning tasks including social networks [1], recommendations [2], and natural language processing [3]. To meet the needs for efficient training, GNN frameworks such as DGL [4] and PyG [5] have been developed, most of which leverage the powerful computing capacity of GPUs. However, the memories of GPUs are usually limited, which makes learning on large-scale graphs difficult.

X. Zhou et al. (Eds.): GPC 2024, LNCS 15225, pp. 69–85, 2026.
https://doi.org/10.1007/978-981-95-1346-8_5

The reason is that the limited GPU memory is not capable of loading the entire graphs and features. Some related full-batch training works [6] are not efficient enough or even feasible.

To scale GNN training on large-scale graphs, some sampling-based mini-batch GNN training systems [7–9] have been designed. Rather than updating GNN models with all vertices of the entire graph, sampling-based GNN training utilizes a portion of nodes in each iteration. Some previous researches [10] have established that mini-batch training converges much faster than traditional full-batch training on huge graphs while maintaining accuracy. Generally, the GNN training process can be divided into three stages: (1) subgraph sampling, (2) feature collection, and (3) model computation. However, the first two stages, especially the feature collection are often the bottlenecks in the training, [11] which usually takes a very high percentage of the whole training time.

To overcome the above problems, some endeavors have been made in recent years. PaGraph [8] adopts a degree-based cache policy to cache the data accessed frequently to the local GPU memory, thus effectively reducing CPU-GPU traffic. However, the cache policy is very simple and ignores the node access characteristics. Moreover, the problem of CPU resource contention still exists. PyTorch-Direct [12] introduces a unified tensor and enables GPU-centric data access, avoiding buffer writing and saving the CPU time of feature collection. However, PyTorch-Direct does not exploit the data locality in GNN training. Furthermore, neither explores the hierarchies of the memory access bandwidth in a CPU-GPU heterogeneous environment.

While the above work can accelerate feature collection in GNN training to a certain extent, some inharmonious phenomena still exist. On one hand, the feature collection still occupies the majority of the execution time. The experiment results reveal that about 55% to 75% of the training time per epoch is spent on the feature collection stage in the aforementioned systems (more details see in the next section). On the other hand, GPU resource utilization is usually low. Our experiments show that the GPU utilization is only around 20%-30% in these systems.

Inspired by the aforementioned issues, to break through the bottleneck of mini-batch GNN training systems, our work focuses on further accelerating the feature collection stage by leveraging some potential features of GNN and GPU resources. Specifically, this paper proposes FCGraph, a novel efficient GNN training strategy leveraging the combination of a feature cache policy of multi-hop (k-hop) neighbors and a hierarchical GPU-centric data access method.

Our main contributions can be summarized as follows:

1. We propose a k-hop neighbors based cache policy by exploiting vertex data access characteristics in k-layer GNN models, which largely reduces redundant vertex accesses and CPU-GPU data transfer overhead.
2. We present a hierarchical GPU-centric data access approach for feature collection by leveraging the cache policy and partitioning node features by access frequency to make better use of GPU. It further explores the locality of access in node feature data and mitigate CPU resource contention.

3. We scale FCGraph to multi-GPU systems equipped with NVLink, in which the memory access hierarchies are explored and the memory bandwidth is better utilized.

2 Background and Motivation

Sampling-based training is the mainstream approach for GNN training. Compared to traditional full-batch GNN training, it utilizes a portion of nodes in each iteration and updates the GNN model per batch. The training workflow of a sampling-based GNN system of DGL is demonstrated in Fig. 1. The graph store is a component that is used to store graph information. The main training workflow is explained as follows:

(1) **Subgraph Sampling.** A subgraph is sampled from the original graph topology by a sampler and transferred to a GPU worker.
(2) **Feature Collection.** Once the GPU worker receives the subgraph information, the features scattered in the CPU memory are gathered from the graph store and then transferred to the GPU worker.
(3) **Model Computation.** Similar to the workflow of DNN, the GPU worker performs the forward computation and backward propagation. And an optimizer (*e.g.*, SGD, Adam) updates the model parameters in each iteration.

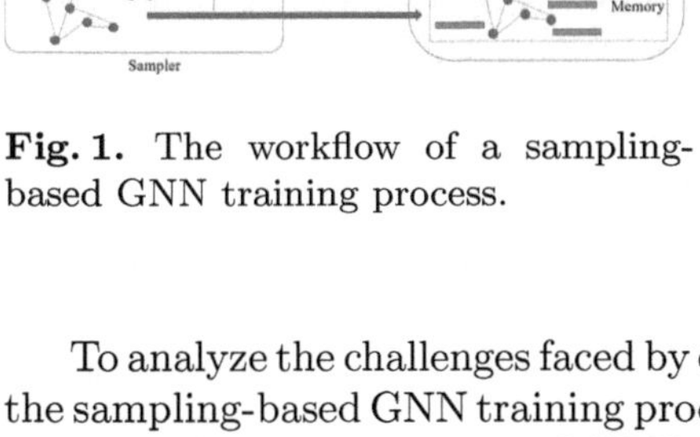

Fig. 1. The workflow of a sampling-based GNN training process.

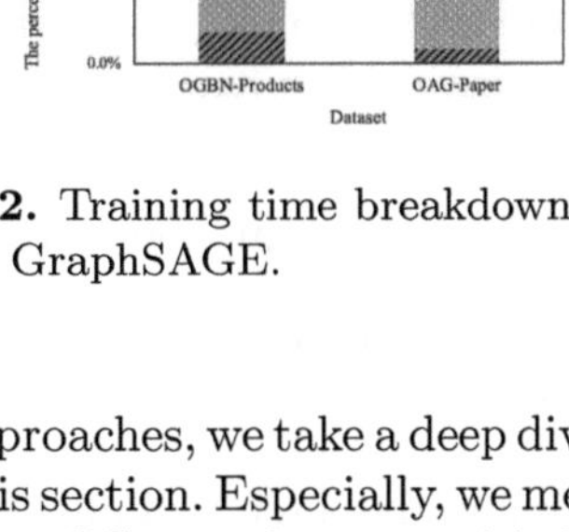

Fig. 2. Training time breakdown for 3-layer GraphSAGE.

To analyze the challenges faced by existing approaches, we take a deep dive into the sampling-based GNN training processes in this section. Especially, we measure end-to-end execution time and GPU utilization on different datasets with 3-layer GraphSAGE [13] on DGL. Some major observations and limitations of existing solutions are discussed below, which motivates us to present FCGraph:

High Proportion of Feature Collection Time. Figure 2 shows the breakdown of the time each stage takes to train GraphSAGE on OGBN-Products [14] and OAG-Paper [15] on a single GPU. Specifically, 55% to 75% of the training time per epoch is spent on feature collection, including feature slicing and CPU-GPU data copy. Additionally, Loading mini-batch data from CPU to GPU through slow PCIe interconnects could be time-consuming.

Massive Redundant Vertex Access. During training, some vertices are loaded multiple times. According to our measurements, the total loading time of corresponding nodes in a training epoch adds up to 3.7 times the sum of the nodes in the original graph. This massive redundant access and feature data transfer severely hurts overall training performance.

Low GPU Resource Utilization. Table 1 shows some GPU utilization when training GraphSAGE on three datasets on a single GPU. We see that the utilization remains low - around 20–30%. The potential power of GPU has been underutilized since CPU is unable to load data from the graph store fast enough. It is worth noting that CPU is suffering resource contention when performing multi-GPU training.

Table 1. GPU utilization when training GraphSAGE on a single GPU.

Datasets	OGBN-Products	Reddit	OGBN-Proteins
GPU Utilization	23%	28%	25%

3 Related Work

Aiming to reduce the amount of intermediate feature transmission in the feature collection stage and alleviate the serious resource contention on the CPU side, many researchers have provided a variety of solutions. Existing works mainly fall into two categories: feature caching and efficient data access patterns.

3.1 GNN Systems Leveraging Feature Caching Strategies

To cut down on CPU-GPU transfers and cross-device communication in feature collection, many GNN-oriented feature caching strategies have been proposed at the system level. These approaches can be divided into two categories: static caching and dynamic caching.

The static caching strategy is to predetermine the node feature data that needs to be cached before training, which can be regarded as a preprocessing step. PaGraph [8] caches frequently accessed data to the local GPU memory and effectively reduces CPU-GPU traffic. While vertex access patterns in GNN workflows are not explored in its cache policy. Moreover, it does not address the CPU contention problem. AliGraph [9] designs a working module that defines an index called l-th importance which comprehensively considers the l-hop out-degree and l-hop in-degree. Nodes are only cached when this metric is greater than a certain threshold. The cache module greatly reduces the communication between partitions. However, AliGraph is a GNN training system optimized only for CPUs and fails to explore the optimization of GNN on GPUs.

In order to achieve a higher cache hit rate during GNN training, some dynamic caching strategies are proposed. BGL [16] integrates a feature cache engine which uses a dynamic caching policy to take the lead in applying a FIFO caching strategy by utilizing the distribution of node features. To further improve the dynamic cache hit rate, BGL fully exploits and utilizes the temporal locality between mini-batches, and introduces Proximity-aware Ordering (PO) during mini-batch preparation. However, the main problem is that the CPU side needs to start and stop the GPU side kernel function frequently, thus massive synchronization overhead will be generated.

3.2 GNN Systems Using Efficient Data Access Patterns

Aside from feature caching issues, researchers found that the problem of CPU data contention is also severe in a distributed environment. In contrast, GPU utilization remains generally low, with abundant parallel computing capacity left untapped. Designing an efficient data access pattern will help further improve the end-to-end performance of GNN training.

P^3 [7] adopts a novel pipelined push-pull parallelism strategy, fusing intra-layer model parallelism and data parallelism to reduce feature collection overheads. However, this hybrid parallel approach brings up additional synchronization overhead.

PyTorch-Direct [12] introduces a unified tensor and enables the GPU to efficiently access corresponding node features scattered in host memory directly without CPU intervention. The GPU-centric data access method avoids buffer writing and saves the CPU time of feature collection. However, this method just puts data on page-locked memory and the data transfer time through PCIe still occupies substantial GNN training time. The memory access hierarchies in a CPU-GPU heterogeneous environment are not fully explored.

4 Methodology

Motivated by the observations mentioned in Sect. 2, we propose FCGraph to lower the cost of CPU-GPU data transfer and efficiently utilize memory bandwidth. It consists of two major modules: a k-hop neighbors based feature cache module and a novel GPU-centric data access method. The former determines which nodes are reserved in local GPU memory and peer GPU memory by counting the number of k-hop neighbors of the target nodes. This cache policy pre-allocates the location of the graph data and aims to reduce CPU-to-GPU data transfer. The latter is based on the former module. By further exploring memory access hierarchies, it utilizes GPU to perform feature collection to fully utilize memory bandwidth and alleviate CPU resource contention. The two major modules are explained as follows.

4.1 *k*-Hop Neighbors Based Feature Cache

As shown in Sect. 2, the feature collection stage occupies the majority of the execution time due to frequent vertex data transfer. Previous research proves that caching part of data on local GPU could effectively reduce CPU-GPU traffic. PaGraph proposes a degree-based cache policy. It claims that a vertex with a higher out-degree is more likely to be sampled in a mini-batch since it becomes an in-neighbor of other vertices with a higher probability. Thus, it selects vertices with a higher degree of priority to fill up the cache room reserved. This static cache policy is simple but can achieve high cache hit ratios. However, it only considers the degree of the node as the caching indicator.

To further lower the cost of CPU-GPU data transfer, a k-hop neighbors based feature cache module is carefully designed in this paper. In the workflow of the mini-batch GNN training, we notice that k-hop neighbors of the target nodes are involved and should be considered in the mini-batch preparation stage. The data loading in sampling-based GNN training workflow is demonstrated in Fig. 3. v_0 is the target node. The nodes of set $\{v_2, v_4\}$ are the sampled one-hop neighbors. The nodes of set $\{v_5, v_{10}, v_{11}\}$ are the sampled two-hop neighbors. Therefore, the corresponding node set in the mini-batch is $\{v_0, v_2, v_4, v_5, v_{10}, v_{11}\}$.

Fig. 3. Data loading in sampling-based GNN training workflow.

More generally, assuming that the node set of target nodes is S and the node set of the sampled k-hop neighbors is N_k, the corresponding node set in a mini-batch can be obtained by $N_1 \cup N_2 \cup ... \cup N_{k-1} \cup N_k \cup S$. When training a k-layer GNN model for a given target node set S, the nodes in the multi-hop neighbor set of S become the nodes of the data flow computation graph in the mini-batch. The data flow computation graph shows the computation dependencies between layers in the GNN training, which is similar to the message flow graph (MFG) in DGL [4]. For the nodes in a certain layer, the features of these nodes will be accessed and collected in an iteration and transmitted to the GPU for calculation. Further analysis shows nodes with larger multi-hop neighbor sets are more likely to be sampled during training. This also increases the probability of their feature data being accessed, collected and transmitted.

Accordingly, the size of the multi-hop neighbor set of a node should determine cache priority when designing the feature cache strategy. Nodes with more multi-hop neighbors have higher access probabilities, making their features better caching candidates. A straightforward idea is to compute the number of multi-hop neighbors of the target nodes and sort nodes descending by this metric. Then, the feature tensors could be organized from hot data to cold data, and the cached part of the features could be reserved in advance.

An example data flow computation graph for training a two-layer GNN model is shown in Fig. 4 to illustrate the node access pattern. We analyze the access probabilities for two nodes, V_1 and V_2. The arrow directions in the figure indicate the calculation directions of the data flow graph constructed in the current GNN training iteration. For V_1, it is the 2-hop neighbor of V_3, V_4, V_6, V_7, and V_8, and the 1-hop neighbor of V_5, V_6, V_9, and V_2. Therefore, V_1 may be sampled when these nodes are trained as target nodes. For V_2, it is the 2-hop neighbor of V_1 and V_7, the 1-hop neighbor of V_3 and V_4. Note that, V_1 has a larger multi-hop neighbor set than V_2, making V_1 more likely to be sampled during GNN training. This also gives V_1 a higher probability of its feature data being accessed, collected, and transmitted.

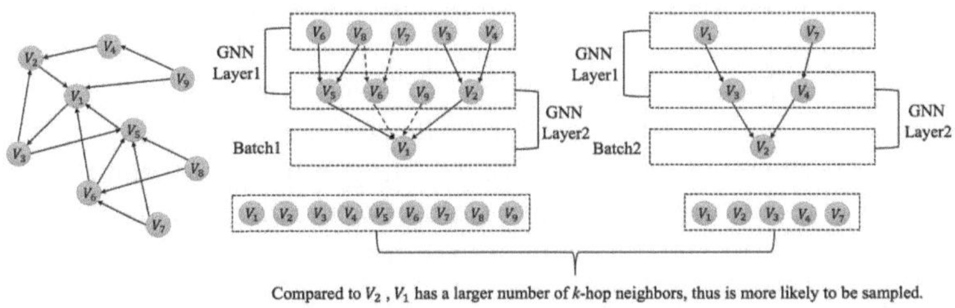

Fig. 4. The node access pattern of the data flow computation graph during training of a two-layer GNN model. (batch size = 1, sampling neighbor = 2)

In tradition, calculating k-hop neighbors of target nodes involves traversing graph g to extract the k-hop subgraph. Assuming that the number of nodes is N, the number of model layers is k, and the average node degree is D, then the time complexity of the recursive algorithm could be $\mathcal{O}(D^k N)$. Clearly, this approach is time-consuming. Instead, we parallelize the algorithm leveraging the built-in *update_all*() function of DGL. The detailed algorithm is presented in Algorithm 1.

In our implementation, the k-hop neighbors of the target node are defined as *kdegree* which is an attribute of g. When the *update_all*() function is executed, the message function *u_add_e*() adds the *kdegree* attribute of the source node to

Algorithm 1: Parallelized k-Hop Neighbors Computation

Input: graph g, number of layers k, device *device*
Output: k-hop neighbors *kdegree*

```
// Save a reversed copy of graph
rev_g = reverse(g).clone();
// Initialize node and edge data
rev_g.ndata['kdegree'] ← 0;
rev_g.edata['kdegree'] ← 1;
rev_g.to(device);
// Repeat k times, then the node data becomes the number of k-hop
   neighbors
for i ← 0 to k − 1 do
  | rev_g.update_all(u_add_e('kdegree','kdegree','m'), sum('m','kdegree'))
end
kdegree = rev_g.ndata['kdegree'].squeeze(dim = 1).int();
return kdegree
```

the *kdegree* of the target node and stores the result in the m attribute of the edge, which is stored in mailbox$['m']$. Subsequently, the aggregation function *sum*() sums the m attributes of all received mailboxes (mailbox$['m']$) and saves the final count to mailbox$['kdegree']$. Formally, this step is very similar to the process of calculating the out-degree of a node. After setting all the node attributes *kdegree* in the reverse graph to 0, and the edge attributes *kdegree* of all nodes in the reverse graph to 1, every time the *update_all*() message is passed once more, the set of neighbors of all nodes in the whole graph is updated globally for each additional hop. So after k calls, the multi-hop neighbors of all nodes in a k-layer GNN are counted. Since *update_all*() parallelizes across vertices, this one-time preprocessing overhead is negligible for training. Our measurements show the cache algorithm execution occupies less than two training epochs across datasets. Details are presented in the experiments section.

4.2 Hierarchical GPU-Centric Data Access

Based on the analysis of the performance bottleneck of GNN training systems in Sect. 2, a novel hierarchical cache-based GPU-centric data access method is proposed to efficiently utilize memory bandwidth and alleviate CPU resource contention. The detailed designs of cache-based GPU-centric data access methods in single-GPU and multi-GPU environments are presented in this section, respectively. Furthermore, the strategy designs in multi-GPU environments are further subdivided according to whether the system is equipped with high-speed NVLink interconnection or not.

Single GPU System. The feature data distribution and access method in a single-GPU system are demonstrated in Fig. 5. Based on the parallelized k-hop neighbors computation algorithm presented above, the features of the nodes with a large number of k-hop neighbors are referred to as hot data. They are allocated in the local GPU memory. Hot data access benefits from the high-speed bandwidth of local GPU memory.

Inversely, the features of the nodes with a small number of k-hop neighbors are referred to as cold data and are allocated in the CPU-pinned memory. For the cold data access, we do away with traditional CPU-centric DMA-based access. In this way, the features scattered in the CPU memory are gathered and written to a buffer, finally transferred to the GPU worker. While, the features gathered in a mini-batch generally range from 256B to 8KB, which is considerably smaller than the large block sizes that enable efficient data transfer in DMA. Moreover, frequent kernel launches by the CPU incur big synchronization overhead, thus declining the performance of the process.

Therefore, a novel GPU-centric data access method is proposed here. The feature collection is performed by GPU without CPU intervention. Inspired by the unified tensor implementation in DGL, we introduce a mixed tensor and design a cache-based GPU-centric data access method. When the mixed tensor is implemented, the method of *append*() of the class of $Torch :: Tensor$ in PyTorch is overridden, and the index operator [] is overloaded. This effectively provides a "fake continuous" data access interface for users, hiding the underlying cross-device physical allocation. As a result, users can utilize the original data indexing scheme without noticing the heterogeneous computing environment. The detailed implementation is based on zero-copy access characteristics of NVIDIA GPUs. In this mode, the CPU memory is page-locked and mapped into the GPU memory space. The CPU memory space can be directly accessed from the GPU, leveraging a unified memory address.

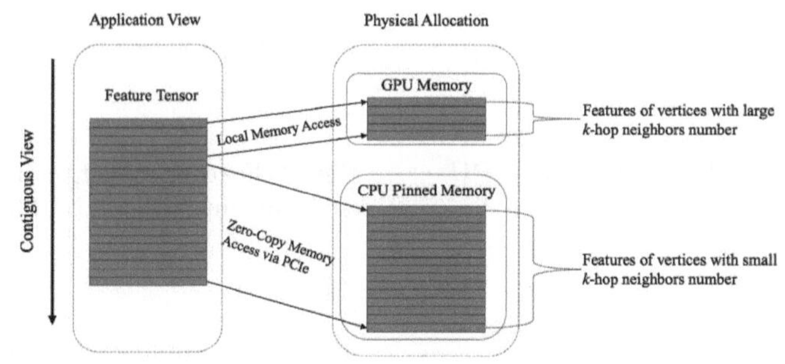

Fig. 5. Feature data distribution and access method overview.

Multi-GPU System. FCGraph can also be scaled to multiple GPUs and further accelerates the GNN training. For a multi-GPU environment, most of the existing GNN training systems ignore the feature of Peer-to-Peer (P2P) data access between GPUs when performing distributed training. In our proposed FCGraph, the memory access hierarchies of the heterogeneous system are further explored and utilized. Specifically, for a multi-GPU heterogeneous system equipped with high-speed NVLink interconnection (such as Tesla V100), the memory access hierarchies of the system are as follows: GPU global memory access bandwidth (about 900 GB/s) is greater than NVLink Interconnect bandwidth (about 300 GB/s) which is greater than PCIe link bandwidth (about 32 GB/s). This hierarchical relationship among the memory access bandwidths within the heterogeneous system forms an important foundation for the cache-based GPU-centric data access method design for the multi-GPU environment described in this section.

The data allocation and access strategy on multi-GPU systems equipped with NVLink are presented in Fig. 6. The feature tensors are partitioned into three parts - those with the top largest k-hop neighbors reside in local GPU memory since they are most likely to be accessed in training, while the next largest ones are placed on peer GPU memory benefiting from the high-speed NVLink bandwidth. The remainder resides in CPU-pinned memory. This tiered placement targets the most common access patterns to local GPU memory first, followed by peer GPUs via NVLink. For the multi-GPU systems without NVLink, the feature collection of FCGraph is demonstrated in Fig. 7. Due to the inability to benefit from high-speed GPUDirect P2P access, we replicate the cache on each GPU, and the overall feature collection workflow of a specific GPU is the same as the design demonstrated in Fig. 5.

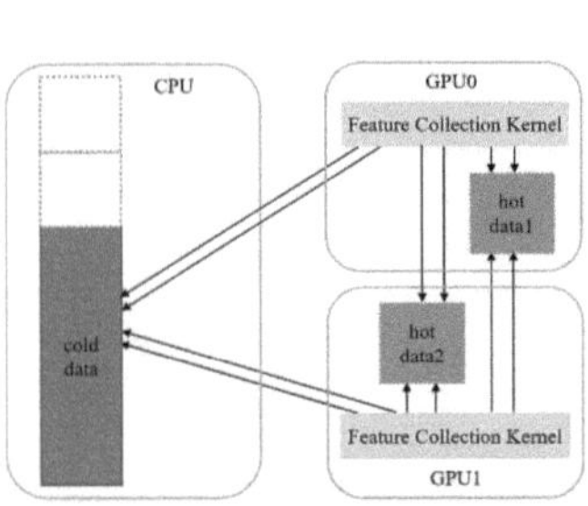

Fig. 6. Feature collection in multi-GPU environment with NVLink.

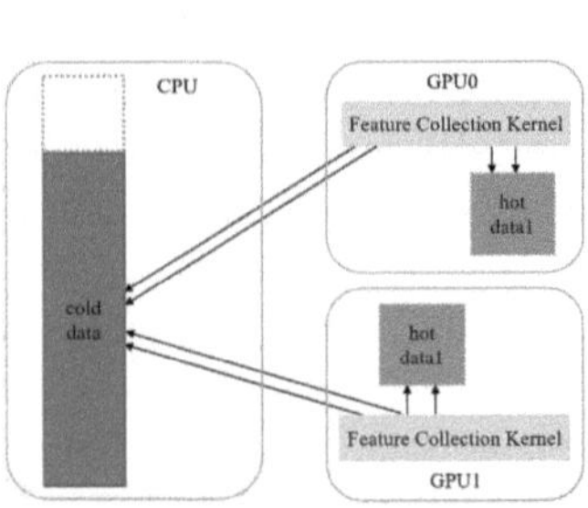

Fig. 7. Feature collection in multi-GPU environment without NVLink.

5 Evaluation

We conduct in-depth evaluation on some representative datasets. We compare the training performance of FCGraph with some mainstream GNN training framework (DGL) [4] and advanced GNN system (PyTorch-Direct) [12]. Specifically, the single-GPU and multi-GPU end-to-end training performance are explored, and the good scalability of FCGraph is shown. Furthermore, we demonstrate that the caching policy of FCGraph outperforms that of PaGraph.

5.1 Experimental Setup

Environments. For evaluation, we deploy the experiments on a multi-GPU machine consisting of 2 Intel Xeon Gold 5117 CPUs and 4 NVIDIA Tesla V100 GPUs (32 GB memory) connected via NVLink. The machine is installed with Ubuntu 18.04, CUDA library v10.2, DGL v0.8.1, PyTorch v1.10.0, and NCCL v2.12.10. For gradient synchronization, we choose the All-Reduce strategy.

Datasets. We use three representative graph datasets listed in Table 2 for evaluation. OGBN-Products represents the co-purchasing relationships of products sold on Amazon. Reddit [13] is a dataset consisting of 232,965 Reddit posts with an average node degree of 492. OGBN-Proteins is a graph in which nodes indicate proteins and edges represent the biological associations between proteins.

Table 2. Datasets in the evaluation (K: Thousand, M: Million).

Dataset	Vertices	Edges	Train Vertices	Features
Reddit	233K	11.6M	153.4K	602
OGBN-Products	2.45M	61.8M	197K	100
OGBN-Proteins	132K	79.1M	86.6K	8

GNN Models. We use two representative GNN models: Graph Convolutional Network (GCN) [12] and GraphSAGE [13] to evaluate FCGraph with detailed configurations in Table 3.

Table 3. Model architectures for evaluation.

Model	Layers	Hidden-dimension	Sampling Fan-out	Batch Size
GraphSAGE	3	256	(15,15,15)	1024
GCN	2	256	(25,10)	1024

Baseline. As aforementioned, DGL and PyTorch-Direct are the two objects for comparison. For the sake of fairness, all of the systems above adopt a GPU sampling strategy. The node feature cache ratio of FCGraph on the three graph datasets is set to 20% during performance testing according to our experimental experience. Moreover, to fairly compare the cache policies of FCGraph and PaGraph, we combine the degree-based cache policy of PaGraph with a GPU-centric data access strategy.

5.2 Single GPU End-to-End Training Performance

We first evaluate the training performance of FCGraph on a single GPU. Figure 8 shows the end-to-end training performance of GraphSAGE on different datasets. Specifically, FCGraph achieves significant speedups of 2.20× for Reddit, 2.67× for OGBN-Products, and 2.72× for OGBN-Proteins, compared to DGL. Moreover, compared to PyTorch-Direct, FCGraph achieves speedups of 1.16×, 1.45×, and 1.26×, respectively. Unlike PyTorch-Direct, FCGraph caches a portion of hot data into local GPU memory, significantly reducing CPU-GPU traffic and further accelerating the feature collection stage.

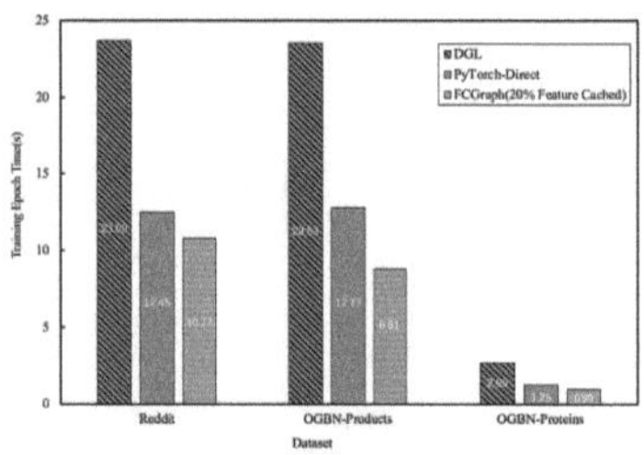

Fig. 8. Training performance on GraphSAGE on a single GPU.

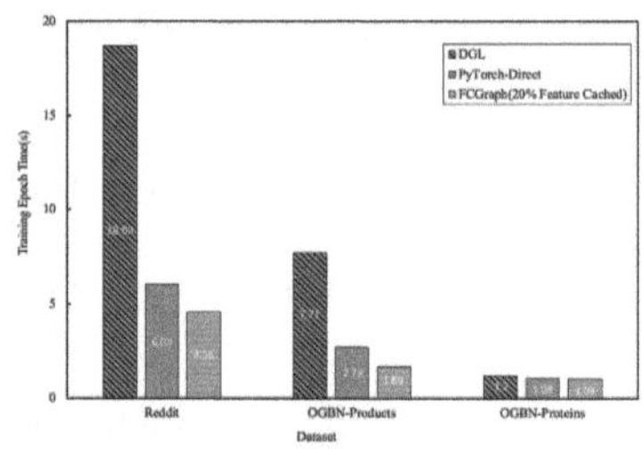

Fig. 9. Training performance on GCN on a single GPU.

Additionally, Fig. 9 shows the training performance of three GNN systems with a two-layer GCN model on the Reddit, OGBN-Products, and OGBN-Proteins datasets. Compared to the three-layer model of GraphSAGE, the two-layer GCN model samples fewer neighbor nodes when aggregating neighbors, resulting in reduced training epoch times. Similar to the training results of GraphSAGE, FCGraph outperforms DGL by speedup ratios of 4.11×, 4.56×, and 1.17×, and PyTorch-Direct by 1.33×, 1.61×, and 1.05×. This highlights the adaptability of our feature caching strategy for GNN models with different layers on different datasets.

5.3 Multi-GPU End-to-End Training Performance

The multi-GPU training performance of DGL, PyTorch-Direct, and FCGraph is evaluated on 2 GPUs and 4 GPUs, respectively. Figure 10 and Fig. 11 show the end-to-end training performance of GraphSAGE and GCN on 2 GPUs for various datasets. It is worth noting that FCGraph, equipped with NVLink, achieves an average speedup of 3.8× and 3.76× compared to DGL on GraphSAGE and GCN, respectively. Additionally, compared to PyTorch-Direct, FCGraph equipped with NVLink achieves an average speedup of 1.67× and 1.59× on GraphSAGE and GCN, respectively. Especially, on OGBN-Products, the speedup can even reach 2.05×. It highlights that FCGraph achieves better performance improvements in multi-GPU environments than single-GPU environments.

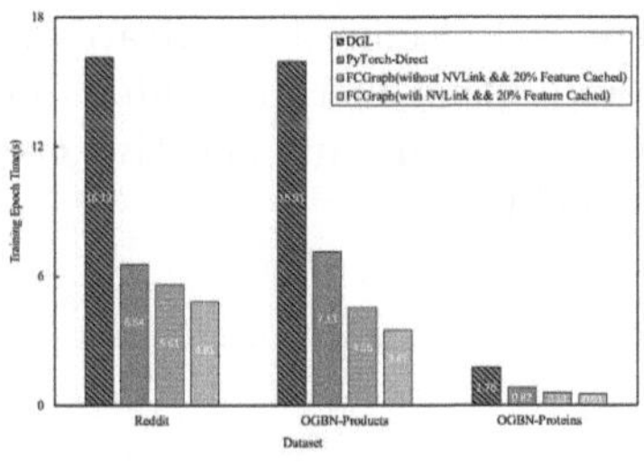

Fig. 10. Training performance on GraphSAGE on 2 GPUs.

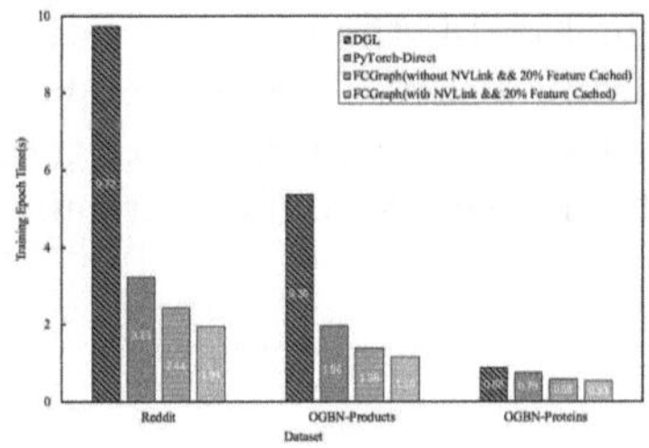

Fig. 11. Training performance on GCN on 2 GPUs.

Figure 12 and Fig. 13 illustrate the end-to-end training performance of GraphSAGE and GCN on 4 GPUs for various datasets. On GraphSAGE, FCGraph equipped with NVLink achieves a speedup of 4.23× compared to DGL and 1.37× compared to PyTorch-Direct. On GCN, it achieves a speedup of 4.69× compared to DGL and 1.5× compared to PyTorch-Direct. It is worth noting that, on OGBN-Products, we see a little decline in performance for FCGraph with NVLink compared to FCGraph without NVLink. The most likely reason is that feature tensor loads are different for each GPU in OGBN-Products when different features are cached in each GPU. The synchronization overhead between the four GPUs is even greater. The data placement strategy of FCGraph in the multi-GPU environment could be further optimized in the future.

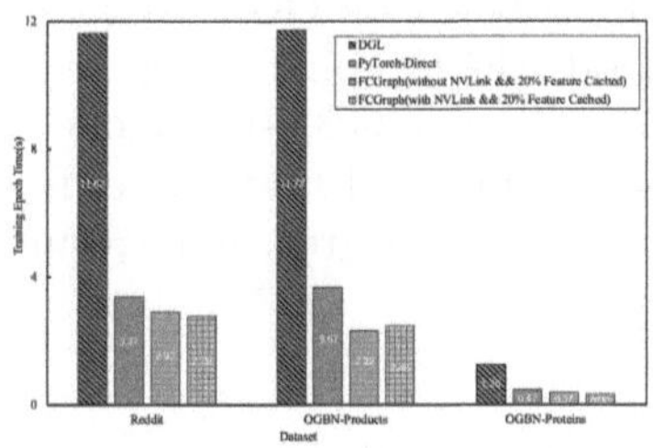

Fig. 12. Training performance on GraphSAGE on 4 GPUs.

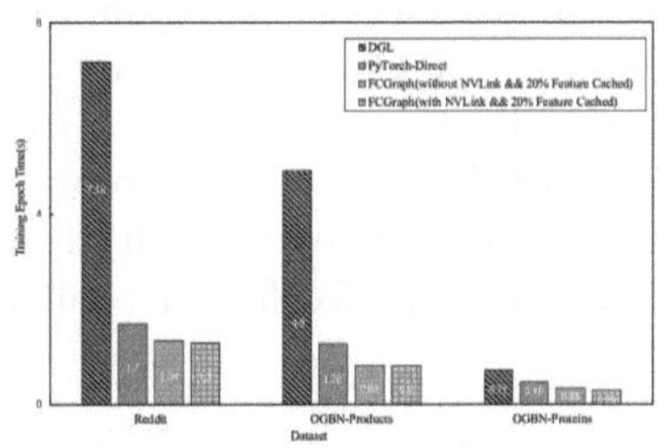

Fig. 13. Training performance on GCN on 4 GPUs.

5.4 Scalability

The scalability of DGL, PyTorch-Direct, and FCGraph is also evaluated. Figure 14 demonstrates the training speedup of GraphSAGE on OGBN-Products relative to single GPU performance of the respective systems on different number of GPUs. That is, each system takes its own end-to-end training time on a single GPU as the benchmark situation. Therefore, when the number of GPUs is 1 in Fig. 14, the speedup ratio of each system is 1. Figure 14 indicates that FCGraph achieves better scalability compared with other methods, which indicates its better exploration of the memory access hierarchy and effective use of the high-speed NVLink interconnection between GPUs.

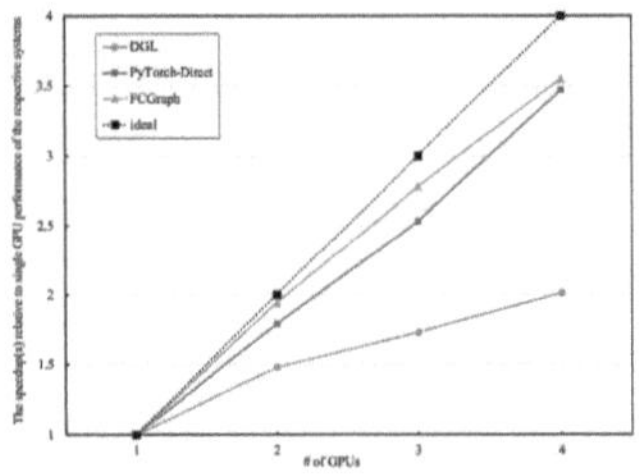

Fig. 14. The scalability of the GNN training systems.

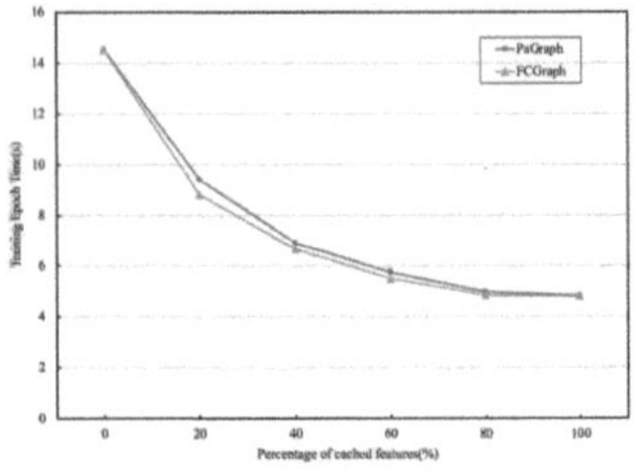

Fig. 15. Training performance varied with cache levels.

5.5 Feature Collection Optimization

To better understand the optimization of our design, the feature collection time per epoch of DGL, PyTorch-Direct, and FCGraph is evaluated on a single GPU. Table 4 lists the execution time of the feature collection stage of GraphSAGE on Reddit, OGBN-Products, and OGBN-Proteins. In general, PyTorch-Direct reduces feature collection time by 60.1% on Reddit, 44.4% on OGBN-Products, and 58.3% on OGBN-Proteins compared to DGL. FCGraph further reduces feature collection time by 69.5% on Reddit, 73.9% on OGBN-Products, and 70.6% on OGBN-Proteins compared to DGL. Keep in mind that the different structural properties of each graph, such as the average degree, may result in varying speedups for feature collection.

Table 4. Feature collection time per epoch on single GPU.

Dataset	DGL	PyTorch-Direct	FCGraph
Reddit	18.42 s	7.34 s (60.1% ↓)	5.62 s (69.5% ↓)
OGBN-Products	18.64 s	10.36 s (44.4% ↓)	4.87 s (73.9% ↓)
OGBN-Proteins	1.97 s	0.82 s (58.3% ↓)	0.58 s (70.6% ↓)

5.6 Effectiveness of Cache Proportion

The effectiveness of the cache proportion is evaluated on a single GPU. Figure 15 illustrates the end-to-end training performance of FCGraph and PaGraph on GraphSAGE with the OGBN-Products dataset. Since the memory access method in PaGraph is still CPU-centric, for a fair comparison, we implement a GPU-centric version of the data access method according to the cache policy of PaGraph. In terms of the cache hit rate comparison between PaGraph and FCGraph, when the cache rate is either 0% or 100%, the epoch time is almost identical since there is minimal difference between the two systems, except for the execution time of the cache algorithm. However, when a portion of the features are cached, we can see that the FCGraph cache policy outperforms the PaGraph cache policy. Thanks to the k-hop neighbors cache policy that FCGraph implements, more node features allocated in the local GPU are accessed in training, and a higher cache hit rate is achieved, leading to a reduction in unnecessary CPU-GPU data transfer and improved end-to-end training performance.

5.7 Execution Time of Cache Algorithm

Table 5 presents the time overhead of the FCGraph cache algorithm and the average time spent per epoch during training when GraphSAGE is trained on various datasets. Except for OGBN-Proteins, where the cache algorithm execution time is approximately equal to 2 epoch training times, the execution time for Reddit and OGBN-Products is less than half of one epoch. For GNN training, which requires several iterations, the execution time of the feature cache algorithm is negligible and can be considered as a preprocessing step.

Generally, we can make a summary that FCGraph can improve training performance over existing GNN training systems of DGL and PyTorch-Direct by 3.72× and 1.46× on average, respectively. Moreover, FCGraph has better scalability when deploying it into multi-GPU environments.

Table 5. Cache algorithm execution time of FCGraph on different datasets and the average time spent per epoch during training.

Dataset	Algorithm Execution Time	Average Epoch Time
Reddit	3.01 s	10.77 s
OGBN-Products	4.14 s	8.81 s
OGBN-Proteins	1.82 s	0.99 s

6 Conclusion

In this paper, we propose FCGraph for efficient GNN training. First, a k-hop neighbors based cache policy is introduced, which achieves higher cache hit ratios compared to advanced existing works and significantly reduces CPU-GPU data transfer. Then, combined with the cache mechanism, a novel GPU-centric data access method is proposed. By exploring memory access hierarchies, we leverage

GPU to perform feature collection to fully utilize memory bandwidth and alleviate CPU-side resource contention. Moreover, we extended our method to work on multi-GPU systems equipped with NVLink, which ensures the scalability of FCGraph. Our evaluation demonstrates that FCGraph outperforms existing GNN training systems such as DGL and PyTorch-Direct obviously. In the future, a challenging and attractive avenue of research is to design an adaptive mode in which feature data can be automatically partitioned and cache size can be dynamically adjusted.

Acknowledgements. This work is supported jointly by the National Key Research and Development Program of China (2022YFB4501400), the National Natural Science Foundation of China (62372199), and the Major Scientific Research Project of Zhejiang Lab (2022PI0AC03).

References

1. Zhang, Y., Gao, S., Pei, J., Huang, H.: Improving social network embedding via new second-order continuous graph neural networks. In: Proceedings of the 28th ACM SIGKDD Conference on Knowledge Discovery and Data Mining, pp. 2515–2523. ACM, Washington, DC, USA (2022)
2. Yuan, K., Liu, G., Wu, J., Xiong, H.: Semantic and structural view fusion modeling for social recommendation. IEEE Trans. Knowl. Data Eng. **35**(11), 11872–11884 (2023)
3. Kumar, V.S., Alemran, A., Karras, D.A., Gupta, S.K., Dixit, C.K., Haralayya, B.: Natural language processing using graph neural network for text classification. In: Proceedings of the 2022 International Conference on Knowledge Engineering and Communication Systems, pp. 1–5. IEEE, Chickballapur, India (2022)
4. Wang, M., et al.: Deep graph library: a graph-centric, highly-performant package for graph neural networks. In: Proceedings of the ICLR Workshop on Representation Learning on Graphs and Manifolds, New Orleans, USA (2019)
5. Fey, M., Lenssen, J.E.: Fast graph representation learning with PyTorch Geometric. In: Proceedings of the ICLR Workshop on Representation Learning on Graphs and Manifolds, New Orleans, USA (2019)
6. Wang, Y., et al.: MGG: accelerating graph neural networks with fine-grained intra-kernel communication-computation pipelining on multi-GPU platforms. In: Proceedings of the 17th USENIX Symposium on Operating Systems Design and Implementation, pp. 779–795. USENIX, BOSTON, MA, USA (2023)
7. Gandhi, S., Iyer, A.P.: P3: distributed deep graph learning at scale. In: Proceedings of the 15th USENIX Symposium on Operating Systems Design and Implementation, pp. 551–568. USENIX, Virtual Event (2021)
8. Lin, Z., Li, C., Miao, Y., Liu, Y., Xu, Y.: PaGraph: scaling GNN training on large graphs via computation-aware caching. In: Proceedings of the 11th ACM Symposium on Cloud Computing, pp. 401–415. ACM, Virtual Event (2020)
9. Yang, H.: AliGraph: a comprehensive graph neural network platform. In: Proceedings of the 25th ACM SIGKDD International Conference on Knowledge Discovery & Data Mining, pp. 3165–3166. ACM, Anchorage, AK, USA (2019)

10. Zheng, D., Song, X., Yang, C., LaSalle, D., Karypis, G.: Distributed hybrid CPU and GPU training for graph neural networks on billion-scale heterogeneous graphs. In: Proceedings of the 28th ACM SIGKDD Conference on Knowledge Discovery and Data Mining, pp. 4582–4591. ACM, Washington, DC, USA (2022)
11. Cao, W., Zheng, C., Yan, Z., Xie, W.: Geometric deep learning: progress, applications and challenges. Sci. China Inf. Sci. **65**(2), 1–3 (2022). https://doi.org/10.1007/s11432-020-3210-2
12. Min, S.W., et al.: Large graph convolutional network training with GPU-oriented data communication architecture. Proc. VLDB Endow. **14**(11), 2087–2100 (2021)
13. Hamilton, W., Ying, Z., Leskovec, J.: Inductive representation learning on large graphs. In: Proceedings of the 31st Conference on Neural Information Processing Systems, pp. 1025–1035. Curran Associates, Inc., Long Beach, CA, USA (2017)
14. Hu, W., et al.: Open Graph Benchmark: datasets for machine learning on graphs. In: Proceedings of the 34th Conference on Neural Information Processing Systems, pp. 22118–22133. MIT Press, Vancouver, Canada (2020)
15. Zhang, F., et al.: OAG: toward linking large-scale heterogeneous entity graphs. In: Proceedings of the 25th ACM SIGKDD International Conference on Knowledge Discovery & Data Mining, pp. 2585–2595. ACM, Anchorage, AK, USA (2019)
16. Liu, T., et al.: BGL: GPU-efficient GNN training by optimizing graph data I/O and preprocessing. In: Proceedings of the 20th USENIX Symposium on Networked Systems Design and Implementation, pp. 103–118. USENIX, BOSTON, MA, USA (2023)

A Survey of Deep Model Compression and Acceleration

Chong Zhang[1], Hongwei Liu[1], Hongzhi Wang[1(✉)], Jiaying Wang[2], Sijia Zheng[2], Xiaoqian Meng[2], and Siyan Zhu[3]

[1] Faculty of Computing, Harbin Institute of Technology, Harbin 150001, China
19B903094@stu.hit.edu.cn, {liuhw,wangzh}@hit.edu.cn
[2] State Grid Electric Power Space Technology Co., Ltd., Nanjing, China
[3] North Automatic Control Technology Research Institute, Taiyuan, China

Abstract. Recently, Deep neural networks (DNNs) have attained remarkable achievements across numerous visual recognition tasks. Nevertheless, the existing deep neural network models are characterized by high computational costs and substantial memory usage, which pose significant barriers to their deployment in devices with limited memory resources or applications with strict latency requirements. Consequently, model compression and acceleration for deep networks without causing notable degradation in model performance is in urgent need. This paper provides a comprehensive review of the recent techniques employed for compacting and accelerating DNN models. From the perspective of model architecture, the main approaches are compact structure design and neural architecture search. From an algorithmic dimension, methods are briefly categorized into static compression methods and dynamic acceleration methods, specifically covering implementation strategies such as model pruning, parameter quantization, low-rank factorization, and knowledge distillation. For each category, we demonstrate the development of mainstream methods as well as the characteristics and advantages of each method. We also provide insightful analysis about the integration of multiple methods, their advantages and drawbacks.

Keywords: Deep Learning · Model Compression · Inference Acceleration

1 Introduction

Deep Convolutional Neural Networks (DCNNs), meticulously designed for processing grid-structured data such as images and audio, have become powerful tools for massive data processing and analysis in the internet era. By leveraging network characteristics such as local connectivity and parameter sharing, convolutional neural networks can significantly reduce the number of required

This work is supported by xx Data Intelligent Management Project(806).

X. Zhou et al. (Eds.): GPC 2024, LNCS 15225, pp. 86–106, 2026.
https://doi.org/10.1007/978-981-95-1346-8_6

parameters while ensuring high performance. However, due to their large number of parameters and computational complexity, deep learning methods have only gradually demonstrated superior performance in numerous application fields since the rapid improvement in the performance of Graphics Processing Units (GPUs) with high-speed parallel computing capabilities. In the field of computer vision, including image classification [1], object detection [2,3], super-resolution analysis [4,5], medical and disease diagnosis [6,7], and in the field of natural language processing, including machine translation [8], semantic segmentation [9,10], and visual question answering [11,12], deep neural networks have achieved outstanding results that surpass traditional algorithms.

The high performance of DCNNs comes at the cost of enormous parameter counts and massive floating-point operations. Take the standard VGG16 [13] model as an example, which has a static parameter size of approximately 300MB. In practical computations, besides static parameters, a large number of intermediate calculation parameters are generated, and parameters need to be continuously read from and written to memory. In terms of computational load, when processing a 224×224 pixel image, the VGG16 model performs approximately 1.5 billion floating-point operations.

Such huge parameter counts and computational loads make it difficult to apply deep convolutional network models in scenarios with limited storage and computing capabilities. For instance, in the field of wearable devices, their compact size imposes extremely limited chip memory space and typically lacks support for high-performance GPUs. However, wearable devices have high requirements for real-time computation and accuracy when running related applications. For intelligent driving, where on-vehicle computing devices also face limitations in memory and computing resources, yet need to quickly and accurately process information such as road conditions. In these scenarios, the application of deep convolutional network models remains highly constrained. Therefore, on the basis of ensuring model performance, compressing and accelerating deep convolutional neural networks has become particularly important. This is a key step to promoting the widespread application of deep convolutional networks in more resource-constrained scenarios.

As deep neural networks continue to evolve, their depth, width, and computational complexity have all increased. However, the expansion of model scale and the increase in the number of parameters have approached a saturation point in terms of improving model performance. LeCun et al. [14] pointed out in their research that the performance of neural networks does not depend on all parameters, but only on part of them. From the perspective of model training, when the number of parameters is too large, it may also cause overfitting. At present, there is no mature and rigorous theoretical explanation for the deterministic relationship between specific network structures and performance, which makes parameter compression and inference acceleration of deep convolutional networks a highly challenging problem.

To address this issue, the main approaches in current research can be categorized as shown in Fig. 1. At the model architecture level, there are two primary

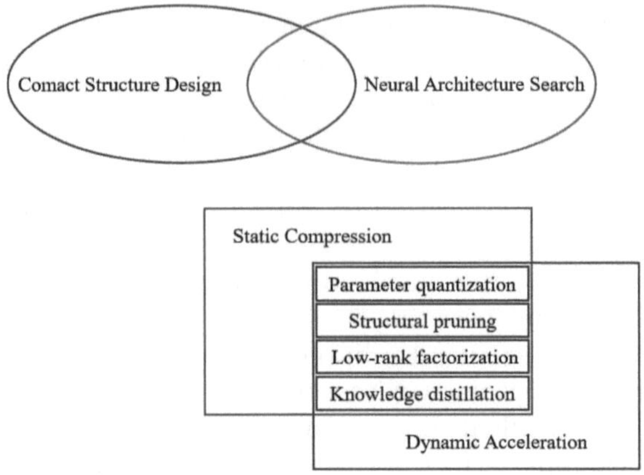

Fig. 1. A summary of compression and acceleration methods for DNN.

solutions: Compact structure design and Neural Architecture Search (NAS). Algorithmically, methods are divided into static compression and dynamic acceleration techniques, including strategies such as model pruning, parameter quantization, low-rank factorization, and knowledge distillation. While research approaches in these fields differ in their objectives, each offers unique advantages and is suitable for specific conditions, contributing to the advancement of parameter compression and inference acceleration in Deep Convolutional Neural Networks (DCNNs).

Table 1. Characteristics of different compression methods

Methods	Parameter Reduction	High Precision	Cross-task	Input-dependent	Special Hardware	Complex Tasks
Structural pruning	✓	Moderate retention				Partial scenarios
Knowledge distillation	✓	✓	✓			✓
Parameter quantization	✓	High bitwidth			✓	Partial scenarios
Low-rank factorization	✓	✓				Partial scenarios
Compact structure design	✓	✓				Partial scenarios
Neural architecture search	✓	✓			Partial	Partial scenarios
Dynamic acceleration	✓	✓		✓	Partial	Partial scenarios

However, these research directions exhibit limitations due to their narrow focus, as summarized in Table 1. They struggle to simultaneously achieve optimal architecture search, maximize model performance, and efficiently utilize training resources. Therefore, rather than pursuing individual methods, the joint optimization of multiple compression and acceleration techniques to balance training/inference efficiency, compression ratios, and performance is of greater value.

In the following sections, we will introduce the characteristics and representative methods of each approach, and analyze their strengths and drawbacks.

2 Knowledge Distillation

Knowledge distillation methods leverage one or more pre-trained complex teacher models to provide simplified knowledge and effective learning strategies to structurally simple student models. In this way, the student models can learn from and draw on the advantages of the teacher models, thereby enhancing their own capabilities to handle complex tasks. From different research perspectives, the existing mainstream knowledge distillation methods are roughly classified into response-based methods, feature-based methods, and relationship-based methods. The approximate classification is shown in Fig. 2.

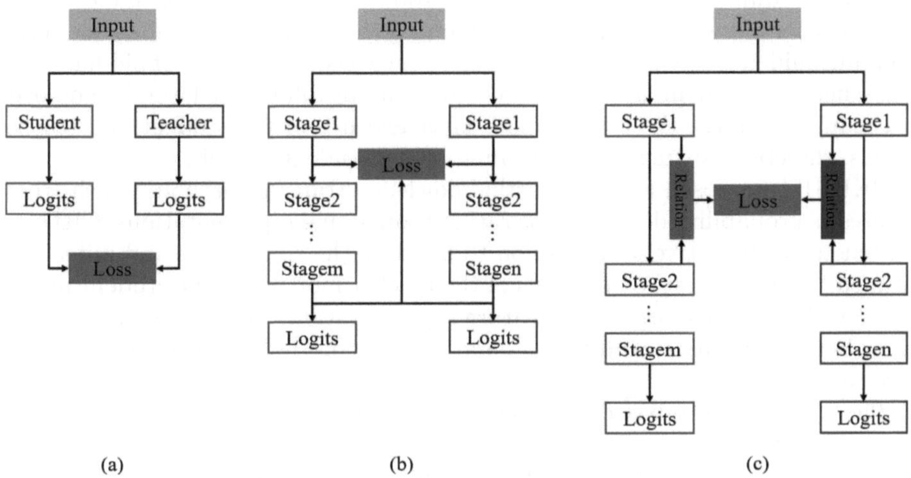

Fig. 2. General classification of knowledge distillation methods.

2.1 Response-Based Methods

Knowledge distillation was first proposed by Hinton [15]. Its core idea is to use the logits output by the teacher network (the output of the last network layer and the input to the softmax layer) as the "soft targets" for student learning, guiding the training of student models together with the real data labels as "hard targets" to achieve knowledge transfer. When calculating the class distribution, the output logits of the teacher model are softened by a temperature parameter T, making the distribution between different classes smoother and facilitating learning by the student model. SpecificallyïijŇfor the final output of student and teacher network a_s and a_t, they are firstly softened by a temperature parameter T before softmax:

$$P_t = softmax(\frac{a_t}{T}), P_s = softmax(\frac{a_s}{T}) \quad (1)$$

then the soft target are combined with the hard target, the distance between student output and the ground truth y_{true}, by a balancing parameter α:

$$L_{KD} = \alpha T^2 D(P_t, P_s) + D(P_s, y_{ture}) \quad (2)$$

2.2 Feature-Based Methods

Although response-based methods significantly enhance the fitting capability of student models, with the development of deep neural networks, solely leveraging the final output to improve student model performance is ultimately limited. To better train deeper student networks, FitNet [16] proposed utilizing knowledge from intermediate network features by introducing the output of hidden layers in the teacher network as "hints" and requiring the student network to use corresponding guided layers to learn these hints. Since the dimensions of hidden layers in teacher and student networks may differ, a convolutional layer is employed as a mapper to align the intermediate layer features of the student and teacher models, thereby assisting in the training of the student model.

PKD [17] proposed a probabilistic knowledge transfer method, which treats features as probability distributions rather than actual representations. CRD [18] constructs a contrastive loss function to measure the similarities and differences between the feature representations of the teacher model and the student model. The model learns discriminative feature representations by maximizing the similarity between positive sample pairs and minimizing the similarity between negative sample pairs.

The above feature-based methods all use single-layer intermediate features for distillation. However, according to the structural characteristics of deep convolutional networks, each network has multiple network stages, and utilizing multi-stage features is expected to achieve better distillation effects. As the distillation method based on attention information of intermediate layer features proposed in [17], it calculates the L2 norm of the feature vector at each position for intermediate layer features and normalizes them to obtain the attention map of the layer, which intuitively displays the image regions that the model focuses on at this layer. VID [18] uses the mutual information technique in information theory to measure the dependence between two random variables. It defines an encoder network to map the features of the student model to a latent space, while defining a decoder network to attempt to reconstruct the features of the teacher model from the latent space. By maximizing the mutual information, the feature representation of the student model can approximate that of the teacher model as closely as possible. The DML [19] method also applies the mutual information approach, enabling multiple untrained student neural networks to collaborate and teach each other, eliminating the dependence on pre-trained teacher networks.

Combining knowledge distillation with GAN (Generative Adversarial Network) could solve task requirements in special scenarios. [20] utilizes GANs to construct an adversarial learning framework, which includes a generator (student model) and a discriminator. The goal of the generator is to produce outputs that can deceive the discriminator, while the discriminator aims to distinguish between the generator's outputs and the real outputs of the teacher model. Through this adversarial training approach, the student model is prompted to learn richer and more accurate knowledge representations. [21] also applies GANs, using randomly generated pseudo-data and a discriminator for training. Aiming at scenarios where training data cannot be obtained due to privacy, security, storage, or other reasons, it achieves data-free knowledge distillation.

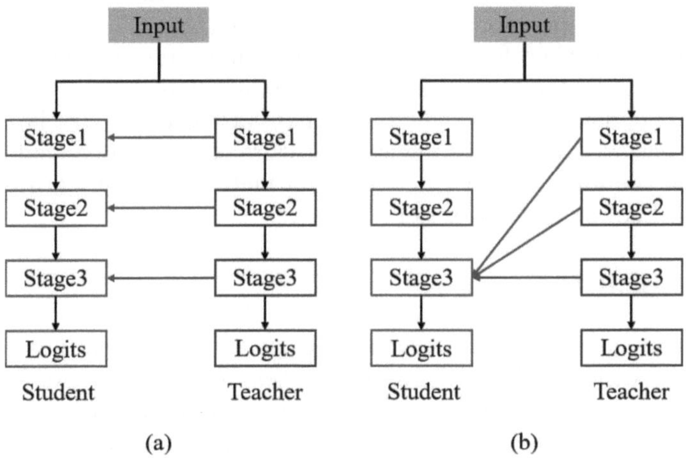

Fig. 3. (a) Knowledge distillation methods transfer knowledge within the same levels. (b) Review mechanism use multiple layers of the teacher to supervise one layer in the student.

Compared with layer-wise feature alignment and knowledge transfer, the Review method proposed in [22] further breaks the limitations between different intermediate layers by introducing a knowledge caching and review approach, as illustrated in Fig. 3. At each stage of student model training, it can review all intermediate features of the teacher model from previous stages, allowing the student model to learn richer and more comprehensive information from teacher knowledge across different stages during training. This helps the student model better capture the intrinsic laws and features of data, avoid falling into local optimal solutions, and improve the model's adaptability to changes in data distribution.

2.3 Relationship-Based Methods

Relationship-based distillation methods further explore the relationships and their changes between different input data and various stages of the model, using these as knowledge for transfer. For example, the similarity-preservation method [25] uses metrics such as cosine similarity to calculate the similarity of sample pairs in the feature spaces of the teacher and student models. Its core lies in enabling the student model to learn the similarity relationships between samples in the teacher model, ensuring that the student's similarity measurements for sample pairs in the feature space are as consistent as possible with those of the teacher model.

The RKD [23] method focuses on the relative relationships between data samples in the feature space and introduces two types of feature relationships based on feature distance and feature angle. By expanding from individual matching to relational matching, it improves the model's generalization ability through learning the relative positions between data points.

The FSP [24] method focuses on the transform of intermediate features generated by the model for input information across different stages. It defines the problem-solving flow through the Gram matrix between different layers. Instead of aligning features directly, it aligns the feature change processes between two stages, using the inner product of the Gram matrices of the features from two layers as the knowledge for transfer.

3 Structural Pruning

In 1989, LeCun et al. [14] proposed the concept of "optimal brain damage". This concept uses second-order derivative information to balance network complexity and training set error, thereby removing unimportant weights in the network. Since then, research on neural network model pruning methods has begun. Before the development of structured and modular deep neural networks, neural networks were generally represented by Multi-Layer Perceptrons (MLPs). Therefore, the pruning of network structures was typically performed at the neuron level. Due to the fine granularity of unstructured pruning, it can maximize the exploitation of redundant connections within the network structure, achieving extreme model compression effects.

For modular network structures like convolutional neural networks (CNNs), pruning can be directly performed on convolution kernels, convolutional filters, or even entire convolutional layers according to the network structure. This approach achieves good compression effects, and the pruned network structure remains regular, without affecting hardware acceleration or deployment.

3.1 Sparsity Methods

From the perspective of sparsification training, the regularization method [25] proposes a general objective formula, group Lasso regularization, which performs sparsification training on the four-dimensional tensor formed by the weights of the convolutional layers in DCNNs to obtain a network structure with sparse model parameters. This allows pruning in any dimension of the four-dimensional tensor. MDP [29] proposes a multi-dimensional pruning method that considers pruning in multiple dimensions simultaneously, and uses collaborative pruning and adaptive pruning strategies to reduce the spatial/temporal and channel redundancies of DCNNs at the same time.

3.2 Channel Pruning Methods

Except for some extreme methods [30] that prune entire layers of the network, general structured pruning methods typically focus on channel pruning. This involves determining the importance of feature channels generated by each convolutional filter in a convolutional layer through evaluation methods, so as to assess the importance of the filters and use it as a basis for pruning.

For example, the HRank [31] method evaluates the importance of convolutional filters by calculating the average rank of feature maps and directly removes filters with low importance. CP [32] refers to the optimization approach of reconstructing feature maps in tensor decomposition, achieving layer-wise pruning by directly minimizing the reconstruction error of output feature maps. BN-Slimming [33] applies L1 regularization to the scaling factors of the batch normalization (BN) layers after convolutional layers, driving the scaling factors toward zero. Since each scaling factor corresponds to a convolutional channel, this allows identifying unimportant channels. This method hardly impairs model performance and even enhances generalization in some cases.

Breaking free from the traditional layer-wise pruning paradigm, Global Ranking [34] obtains global importance ranking metrics for all convolutional filters in the network through a learning-based approach. By comprehensively evaluating the parameters of the entire model, it determines which parameters are more critical to the model's performance and which can be safely removed or compressed. Another global pruning method, GVTP [35], assesses the significance of model parameters using Hessian matrices to achieve global pruning.

3.3 Optimization-Based Pruning Methods

Structured pruning can also be viewed as an optimization problem, where the optimal network structure is obtained under given constraints on model parameters and within a defined structural search space. To start with, [36] transforms the pruning problem into a convex optimization problem and introduces a performance guarantee mechanism, ensuring high performance of the pruned model while achieving compression. Thinet [37] formulates filter pruning as an optimization problem, determining filter importance by evaluating the reconstruction error of input feature maps for the next layer. [38] develops a differentiable pruning criterion sampler, allowing the network to automatically learn filter pruning criteria for different layers. By defining a joint loss function that includes both pruning loss and classification loss, it optimizes network parameters and pruning criteria simultaneously. The pruning loss guides the network to learn appropriate filter importance metrics, while the classification loss maintains good classification performance during pruning.

DMC [39] proposes a model compression framework based on discrete optimization, transforming the problem into an optimization task in discrete space. Considering multiple resource constraints (e.g., memory, computation), it searches for optimal solutions in the discrete model structure and parameter space to achieve compression. Similar to the gating parameter method in

[40], AAF [41] introduces residual gating structures in the network, dynamically controlling channel retention or pruning by learning importance scores for each channel. Unlike [40], AAF does not solely rely on channel importance for pruning. Instead, it comprehensively considers hardware-specific factors (e.g., computational load and memory access overhead) to prioritize pruning channels that contribute more to acceleration with minimal impact on model accuracy, thus improving inference speed while minimizing performance degradation.

3.4 Structured Pruning Combined with Other Methods

Channel pruning methods can also achieve pruning goals when combined with other learning methods.

Combined with automated machine learning methods, AMC [42] uses reinforcement learning to provide model compression strategies, transforming the model compression problem into a reinforcement learning task. An agent learns to determine the optimal compression strategy for each layer of the neural network, achieving an optimal trade-off between model size, speed, and accuracy.

Combined with GAN, The GAL [43] method proposes a pruning strategy based on generative adversarial learning to automatically search for the optimal network structure. The pruned network is treated as the generator, whose output features are labeled "fake" by the discriminator, while the pre-trained model serves as the baseline, with its output features labeled "real". Regularization constraints are introduced to make the discriminator guide the generator to learn the distribution of the baseline outputs while sparsifying the soft masks in the generator to achieve pruning.

Combined with time-frequency transform methods, CNNpack [44] uses the Discrete Cosine Transform (DCT) to convert convolutional filter weights to the frequency domain and compresses the convolutional neural network in the frequency domain. Since learned convolutional filter weights are typically smooth and low-frequency, most weights are concentrated in the low-frequency components in the frequency domain, enabling the differentiation of importance among different convolutional filters. Combined with neural architecture search methods, NAP [45] performs gradient-based neural architecture search while pruning synchronously during the search process, directly yielding lightweight network architectures while reducing the search space.

4 Compact Structure Design and Network Architecture Search

4.1 Compact Structure Design

From the perspective of network structure design, constructing lightweight network architectures that are compact, fast-responsive, and highly accurate can be achieved through carefully designed more efficient forms of convolutional computation and feature extraction methods [46,47]. For example, SqueezeNet [48] replaces traditional convolutional layers with squeeze layers and expand layers.

It uses smaller 1×1 convolution kernels to reduce channel feature computation and ensures feature diversity through feature concatenation of 1×1 and 3×3 convolutions.

MobileNet [49], a lightweight model based on depthwise separable convolutions, decomposes traditional convolutional computation into two steps: depthwise convolution (for spatial feature extraction) and pointwise convolution (for channel-wise feature mixing). This innovative design significantly reduces the computational complexity of convolution operations, enabling the model to operate efficiently in resource-constrained environments. MobileNetV2 [50] further introduces linear bottleneck units and inverted residual modules based on depthwise separable convolutions. By avoiding information loss caused by non-linear activation functions in low-dimensional spaces, it enhances the model's expressive power.

ShuffleNet [51] ingeniously combines pointwise convolution with group convolution, significantly reducing the computational complexity of convolution operations. However, group convolution can lead to poor channel information flow during computation. To effectively address this drawback, researchers proposed the channel shuffle operation. By shuffling and recombining the channels of different convolution groups, this operation allows information to flow again between groups, greatly enhancing the interactivity of feature information.

EfficientNet [52] proposes a compound scaling method by studying the intrinsic relationships among network depth, width, and input image resolution. With this method, it is possible to accurately find the optimal balance between model efficiency and performance under different resource constraints, ensuring that the model achieves the best state in resource utilization and task performance.

4.2 Network Architecture Search

NAS (Neural Architecture Search) is a category of automatic search algorithms whose core objective is to accurately explore network architectures that meet specific requirements within a given search space. In the application context of model compression, these requirements typically include constraints on performance, number of parameters, computational volume, and running time. The development of NAS methods mainly revolves around three key elements: optimizing the search space, search strategy, and evaluating.

Regarding the optimization of search spaces, NAS methods generally face challenges such as excessively large search spaces, insufficient representational capabilities of search spaces, and a lack of transferability in the discovered architectures. To address these challenges, current research has explored a series of effective solutions, including unitized design of search spaces, building supernetworks that encompass all possible architectures, and transferring search results from small to large search spaces. For example, [53] proposes a transferable architecture search method that decomposes neural network architectures into multiple repeating units. By searching for universal network architecture units on small-scale datasets, these units can then be stacked into larger networks for large-scale image recognition tasks. One-Shot NAS [54] adopts the super-network concept, constructing different architectures using parameter-sharing mechanisms. During the search process, it rapidly evaluates the performance of various architectures by training and sampling from the super-network.

From the perspective of search strategies, NAS can leverage traditional methods such as random search [55] and gradient-based search [56,57], or combine with advanced techniques like reinforcement learning [58] and evolutionary algorithms [59] to explore superior architectures.

For instance, [56] introduces a differentiable architecture search (DARTS) method that transforms the discrete architecture search problem into a continuous optimization task. By introducing architecture parameters into a super-network, it jointly optimizes both model weights and architecture configurations using gradient descent. MNas [60] addresses hardware heterogeneity in mobile devices by proposing a platform-aware multi-objective NAS approach. This method simultaneously optimizes model accuracy and inference latency (or other resource metrics) on target hardware. Through a tailored multi-objective optimization strategy, it automatically discovers neural architectures that balance efficiency and accuracy for mobile deployment.

The search efficiency of architecture search methods is significantly constrained by evaluation approaches. Early NAS methods required individual validation for each candidate architecture, which was extremely time-consuming. To address redundant computations, methods like [54,61,62] adopted parameter-sharing strategies. OneNas [61] leverages prior knowledge by designing appropriate prior distributions to guide the search direction. This prior knowledge can stem from insights into neural network structures, existing experiences, or knowledge from related domains, optimizing the search process using gradient-based methods. ENAS [62] represents the search space as a directed acyclic graph (DAG) that superimposes all possible sub-models. By enabling weight sharing among sub-models, it avoids the inefficient process of training each sub-model from scratch to convergence. NWOT [63] introduces a kernel matrix-based scoring method that assesses the overlap of activation values generated by different data inputs in a mini-batch. This allows partial prediction of training accuracy from the network's initial state without actual training.

To enhance search efficiency, methods such as DisWOT [64] and DNA [65] introduce novel training-free evaluation metrics that significantly reduce computational overhead and time costs during architecture search: DisWOT posits that the optimal student model should minimize semantic divergence from the teacher model while maximizing relational similarity. Unlike traditional NAS approaches that require training and evaluating numerous candidate architectures, DisWOT analyzes feature distributions and structural properties between teacher and student networks to enable efficient architecture evaluation without training. DNA proposes a hybrid approach combining knowledge distillation and modular supervision. It partitions the NAS search space into blocks and leverages "neural architecture DNA" extracted from the teacher model as supervision to guide block-wise architecture search, dramatically improving NAS efficacy.

5 Quantization and Low-Rank Factorization

5.1 Parameter Quantization

With the continuous growth in the scale of deep learning models, parameter quantization has garnered extensive attention as an effective method to reduce model storage and computational costs.

In the exploration of quantization methods, early works such as the deep compression framework proposed by Han et al. [66] compressed DNNs through three steps: pruning, quantization, and Huffman coding, enabling AlexNet to be compressed 35 times without accuracy degradation. Zhou et al. [67] introduced a progressive quantization scheme, which first performs grouped quantization, freezes the quantized parts, and trains the unquantized parts, repeating this process until all weights are quantized. Binarized networks reduce accuracy loss caused by quantization by quantizing weights into specific values, allowing computations to be completed via simple shift operations on hardware. Inspired by this, numerous optimization schemes for binarized neural networks have emerged. For example, [68] uses scaling factors for binary parameters to more accurately approximate original floating-point values, minimizing accuracy loss. XNOR-Net [69] replaces the single-step approximation with recursive approximation based on quantization residuals, further reducing quantization errors.

In the field of post-training quantization, significant achievements have also been made. Markus et al. proposed a data-free quantization scheme (DFQ) [70], which eliminates the need for a calibration dataset to fine-tune quantization parameters. By employing weight equalization and bias correction, DFQ enhances the quantization accuracy of networks, enabling 8-bit data-free quantization of both network weights and activations. This approach addresses the challenge of limited data availability in practical deployment scenarios, providing a more flexible quantization solution for model compression. The 4-bit post-training quantization scheme ACIQ [71] (Analytical Clipping for Integer Quantization) improves accuracy by solving for the optimal clipping value and employs a bias-correction method.

To address the significant accuracy loss under extremely low quantization bitwidths, Fang et al. proposed a piecewise linear quantization scheme, PWLQ [72]. By searching for optimal segment breakpoints, PWLQ achieves post-training quantization with only minimal calibration data. However, this method requires specialized hardware support.

In addition to quantization methods themselves, there has been growing research on model compression that integrates other techniques. For example, FSNet [73] introduces a differentiable parameter-sharing framework for model compression. This approach represents all convolutional kernels in a layer using a compact set of "filter summaries", drastically reducing the number of unique parameters. Critically, it can be combined with parameter quantization to achieve even higher compression rates.

5.2 Low-Rank Factorization

Low-rank decomposition is essentially a matrix decomposition technique, which aims to factorize a matrix into the product of two or more matrices. This process compresses high-dimensional data into low-dimensional representations, thereby reducing the number of parameters.

Early research efforts primarily focused on extending two-dimensional matrix decomposition methods to high-dimensional tensor decomposition for accelerating and compressing convolutional networks. Methods involved include CP decomposition, Tucker decomposition, Tensor Train decomposition [74], and Block Term decomposition [75], among others. These approaches laid the foundation for subsequent research and demonstrated that tensor decomposition can effectively reduce the parameter count and computational complexity of convolutional kernels.

Further, MicroNet [76] proposed a lightweight network for extremely low-computation-volume scenarios. It decomposes the original convolution into multiple small convolutions through Micro-Factorized convolution, and uses low-rank approximation to reduce the number of input-output connections without changing the connectivity. Meanwhile, it combines with the Dynamic Shift-Max activation function to enhance the network's nonlinear ability, achieving a good classification accuracy under low-computation conditions. Through Singular Value Decomposition (SVD), [77] has found that the model weights in the Transformer structure are not low-rank, while the features (i.e., activations) are low-rank. Based on this, an adaptive search method has been designed to determine the compressed model architecture.

6 Dynamic Inference Acceleration

Static compression methods mainly focus on optimizing network structures and expressive capabilities, primarily applied during the training phase of deep convolutional networks. These schemes reduce both the parameter count and computational load of network models. Although they can achieve significant compression effects, they inevitably limit the upper bound of the network's expressive capabilities.

In contrast, dynamic inference acceleration algorithms focus on capturing differences in input data while preserving all model parameters intact. By dynamically matching optimal computational paths for different inputs and flexibly adjusting the computation flow based on the real-time state of the neural network during runtime, these algorithms achieve efficient acceleration. They can significantly enhance inference efficiency without compromising the model's expressive capabilities, making them particularly suitable for real-time inference scenarios in resource-constrained environments.

Inference acceleration methods can be viewed as inference path selection for different inputs or dynamic model sparsification to some extent. From the perspective of path selection, the RNP [78] method adopts a bottom-up, layer-by-layer selection approach, modeling the inference path selection problem as a

Markov decision process and training it using reinforcement learning. It evaluates the importance of each convolution kernel based on different input samples and performs channel-level pruning.

From the perspective of dynamic sparsification, [79] treats each neuron or connection in the model as a decision variable. Based on the characteristics of input instances, it assigns a sparsity probability to each neuron or connection, indicating the likelihood of being pruned for the current instance. In this way, the model can automatically adjust its computational graph for different input instances, retaining only the neurons and connections that are important for the current input.

Manifold [80] dynamically prunes by mining the manifold information of input instances, analyzing the sample space from two aspects: complexity and similarity. By calculating the similarity between input features and instance saliency vectors, it ensures that similar feature instances have similar saliency vectors, thereby preserving the manifold structure of the data, avoiding pruning from damaging the internal relationships between input instances, and achieving effective removal of redundant filters. [81] designs a novel Dynamic Processing Unit (DPU) based on dynamic pruning methods. During the pruning training iterations, weights are no longer directly pruned or set to 0 but are managed by the DPU, which decides whether they participate in subsequent network forward propagation. This approach retains the flexibility of weights and avoids errors in the intermediate process of dynamic pruning.

From the perspective of interpretability, InterDNP [82] proposes a dynamic feature selection method, equipping each convolutional layer with a computing unit. This unit stores m masks as alternative paths. For different inputs, it calculates the probabilities corresponding to these m masks and selects the mask with the highest probability for channel convolution computation, thereby visualizing the network's decision paths to improve model interpretability.

Dynamic pruning methods can use convolutional filters as pruning units or entire convolutional layers as pruning units. Skipnet [83] uses gating units to dynamically select which layers of the convolutional neural network to skip during inference. The gating network maps the output of the previous layer or group of layers to binary decisions to determine whether to execute the next layer or group of layers.

In addition to dynamically selecting inference paths, another more intuitive and easy-to-implement approach is to use early exit strategies, which can timely stop subsequent computations based on the complexity of inputs and the computations already completed, thus achieving effective allocation of computational resources. For exampleïijŇBranchynet [84] introduces a side-branch mechanism into traditional deep neural networks, where each side branch can serve as a potential "exit point". These side-branch classifiers allow the model to decide whether to continue processing through the main network or directly output results from the processed part according to the complexity of input samples. ACT [85] proposes an Adaptive Computation Time (ACT) algorithm, which enables recurrent neural networks (RNNs) to learn how many computational steps to take between receiving inputs and generating outputs. By adding a Sigmoid stopping unit to the network output, the probability of continuing computation is determined by the activation level of this unit. Furthermore, the SACT [86] method applies the Adaptive Computation Time (ACT) mechanism suitable for recurrent neural networks to residual networks, adaptively adjusting the amount of computation at different spatial locations. This allows the network to dynamically determine the number of layers to execute for each spatial location according to the feature complexity of different regions in an image.

Dynamic inference acceleration techniques are highly compatible with traditional model compression strategies, forming a complementary synergy: During the training phase, methods such as pruning and quantization are used to reduce network scale. During the inference phase, dynamic acceleration techniques further optimize computational resource allocation. This combined strategy of "structural optimization in training + dynamic acceleration in inference" achieves more significant model compression and acceleration effects, providing strong support for the efficient deployment of deep learning models in resource-constrained scenarios.

7 Discussion

In this paper, we provided a survey on deep model compression and acceleration, introduced several lines of work to compress model parameters and FLOPs statically, and accelerate model inference dynamicallyïijŇincluding parameter quantization, low-rank decomposition, model pruning, lightweight network design, neural architecture search, knowledge distillation, and dynamic inference acceleration. Here we analysis the advantages and disadvantages of each method.

- Model pruning can achieve significant model compression. However, as pruning essentially involves optimizing the overall network structure, completely identical pruning strategies cannot generally be applied to network models with different structural characteristics, necessitating customized design. Additionally, manual adjustment of pruning thresholds and repeated fine-tuning of parameters are required, making the implementation process cumbersome.
- The advantages of knowledge distillation lie in its ability to effectively lightweight models, utilize knowledge from pre-trained models, offer high flexibility and strong compatibility (allowing integration with other compression techniques), and maintain relatively good accuracy. Its disadvantages include reliance on high-performance teacher models, high computational costs, an upper limit on student model performance, and sensitivity to hyperparameters, leading to complex parameter tuning.
- Compact architecture design can significantly reduce the number of FLOPs. However, due to the unique design of lightweight convolutional kernels, the compressed models are extremely compact, making it difficult to comprehensively apply other compression techniques to further improve performance. Additionally, compared with large and deep original models, networks

designed with lightweight convolutional kernels have limited capacity, resulting in slightly weaker generalization capabilities of the learned features during training.

- The NAS method can automatically explore high-performance network architectures, breaking through the limitations of manual design. It integrates multiple optimization strategies to adapt to diverse conditions and synergizes with other compression techniques. However, it has drawbacks such as excessive computational power and time consumption during the search process, extremely high resource requirements, and poor interpretability of the discovered architectures.
- Parameter quantization method reduce model parameters to a certain extent, not restricted by model structure. However, the theoretical compression effect often requires the support of special hardware and quantized models are difficult to train and suffer from severe accuracy loss. Nevertheless, quantization has good compatibility and is often combined with other compression methods in practical applications.
- Low-rank decomposition method reduce model parameters and computational complexity through tensor dimensionality reduction, adapt to multiple network architectures, and cause minimal changes to the model. However, decomposition faces complex calculations, time-consuming training, accuracy degradation, and limited standalone compression effects, often requiring collaboration with other techniques.
- Dynamic inference acceleration methods can dynamically adjust computational paths to reduce redundancy, improve efficiency, and adapt to uneven data and real-time scenarios. However, they require additional mechanisms that increase complexity, rely on data and hardware feedback, have poor cross-platform compatibility, and their acceleration effects are constrained by the operating environment.

The above methods are just an overview of current popular techniques. With the increasing interest in this field from both academia and industry, a large number of new methods are being continuously explored. In the future, it is hoped that more combined approaches can integrate the strengths of various compression and inference techniques to achieve better and faster inference results with fewer computational resources.

References

1. Krizhevsky, A., Sutskever, I., Hinton, G.E.: Imagenet classification with deep convolutional neural networks. Commun. ACM **60**(6), 84–90 (2017). https://doi.org/10.1145/3065386
2. Liu, W., et al.: SSD: single shot multibox detector. In: Leibe, B., Matas, J., Sebe, N., Welling, M. (eds.) ECCV 2016. LNCS, vol. 9905, pp. 21–37. Springer, Cham (2016). https://doi.org/10.1007/978-3-319-46448-0_2
3. Ren, S., He, K., Girshick, R., et al.: Faster R-CNN: towards real-time object detection with region proposal networks. In: Advances in Neural Information Processing Systems, vol. 28 (2015)

4. Kong, X., Zhao, H., Qiao, Y., et al.: Classsr: a general framework to accelerate super-resolution networks by data characteristic. In: IEEE Conference on Computer Vision and Pattern Recognition, CVPR, pp. 12016–12025 (2021). https://doi.org/10.1109/CVPR46437.2021.01184
5. Dong, C., Loy, C.C., He, K., et al.: Image super-resolution using deep convolutional networks. IEEE Trans. Pattern Anal. Mach. Intell. **38**(2), 295–307 (2016). https://doi.org/10.1109/TPAMI.2015.2439281
6. Yin, H., Mukadam, B., Dai, X., et al.: Diabdeep: pervasive diabetes diagnosis based on wearable medical sensors and efficient neural networks. IEEE Trans. Emerg. Top. Comput. **9**(3), 1139–1150 (2021). https://doi.org/10.1109/TETC.2019.2958946
7. Niyaz, U., Bathula, D.R.: Augmenting knowledge distillation with peer-topeer mutual learning for model compression. In: International Symposium on Biomedical Imaging (ISBI), pp. 1–4 (2022). https://doi.org/10.1109/ISBI52829.2022.9761511
8. Sutskever, I., Vinyals, O., Le, Q.V.: Sequence to sequence learning with neural networks. In: Proceedings of the 28th International Conference on Neural Information Processing Systems, vol. 2, pp. 3104–3112 (2014)
9. Liu, J., Zhang, F., Zhou, Z., et al.: Bfmnet: bilateral feature fusion network with multi-scale context aggregation for real-time semantic segmentation. Neurocomputing **521**, 27–40 (2023). https://doi.org/10.1016/j.neucom.2022.11.0.84
10. Zhang, B., Wang, Z., Ling, Y., et al.: Mx2m: masked cross-modality modeling in domain adaptation for 3D semantic segmentation. In: Conference on Artificial Intelligence, (AAAI), pp. 3401–3409 (2023). https://doi.org/10.1609/AAAI.V37I3.25448
11. Antol, S., Agrawal, A., Lu, J., et al.: VQA: visual question answering. In: IEEE International Conference on Computer Vision (ICCV), pp. 2425–2433 (2015). https://doi.org/10.1109/ICCV.2015.279
12. Noh, H., Seo, P.H., Han, B.: Image question answering using convolutional neural network with dynamic parameter prediction. In: IEEE Conference on Computer Vision and Pattern Recognition (CVPR), pp. 30–38 (2016). https://doi.org/10.1109/CVPR.2016.11
13. Simonyan, K., Zisserman, A.: Very deep convolutional networks for largescale image recognition. In: International Conference on Learning Representations, ICLR (2015). http://arxiv.org/abs/1409.1556
14. Lecun, Y., Denker, J.S., Solla, S.A.: Optimal brain damage. In: Advances in Neural Information Processing Systems, pp. 598–605 (1989)
15. Hinton, G., Vinyals, O., Dean, J.: Distilling the knowledge in a neural network (2015). arXiv:1503.02531
16. Romero, A., Ballas, N., Kahou, S.E., et al.: Fitnets: hints for thin deep nets. In: International Conference on Learning Representations (2015)
17. Passalis, N., Tzelepi, M., Tefas, A.: Probabilistic knowledge transfer for lightweight deep representation learning. IEEE Trans. Neural Netw. Learn. Syst. **32**(5), 2030–2039 (2021). https://doi.org/10.1109/TNNLS.2020.2995884
18. Tian, Y., Krishnan, D., Isola, P.: Contrastive representation distillation. In: International Conference on Learning Representations (2020)
19. Zagoruyko, S., Komodakis, N.: Paying more attention to attention: improving the performance of convolutional neural networks via attention transfer. In: International Conference on Learning Representations (2017)

20. Ahn, S., Hu, S.X., Damianou, A., et al.: Variational information distillation for knowledge transfer. In: Proceedings of the IEEE/CVF Conference on Computer Vision and Pattern Recognition (CVPR) (2019)
21. Zhang, Y., Xiang, T., Hospedales, T.M., et al.: Deep mutual learning. In: Proceedings of the IEEE Conference on Computer Vision and Pattern Recognition (CVPR) (2018)
22. Wang, Y., Xu, C., Xu, C., et al.: Adversarial learning of portable student networks. In: Proceedings of the Conference on Artificial Intelligence, (AAAI), pp. 4260–4267 (2018). https://doi.org/10.1609/AAAI.V32I1.11667
23. Chen, H., Wang, Y., Xu, C., et al.: Data-free learning of student networks. In: IEEE/CVF International Conference on Computer Vision (ICCV), pp. 3513–3521 (2019). https://doi.org/10.1109/ICCV.2019.00361
24. Chen, P., Liu, S., Zhao, H., et al.: Distilling knowledge via knowledge review. In: IEEE Conference on Computer Vision and Pattern Recognition, pp. 5008–5017 (2021)
25. Tung, F., Mori, G.: Similarity-preserving knowledge distillation. In: IEEE/CVF International Conference on Computer Vision (ICCV), pp. 1365–1374 (2019). https://doi.org/10.1109/ICCV.2019.00145
26. Park, W., Kim, D., Lu, Y., et al.: Relational knowledge distillation. In: Proceedings of the IEEE/CVF Conference on Computer Vision and Pattern Recognition (CVPR) (2019)
27. Yim, J., Joo, D., Bae, J.H., et al.: A gift from knowledge distillation: fast optimization, network minimization and transfer learning. In: IEEE Conference on Computer Vision and Pattern Recognition, pp. 7130–7138 (2017)
28. Wen, W., Wu, C., Wang, Y., et al.: Learning structured sparsity in deep neural networks. In: Proceedings of the 30th International Conference on Neural Information Processing Systems, pp. 2082–2090 (2016)
29. Guo, J., Ouyang, W., Xu, D.: Multi-dimensional pruning: a unified framework for model compression. In: 2020 IEEE/CVF Conference on Computer Vision and Pattern Recognition (CVPR), pp. 1505–1514 (2020). https://doi.org/10.1109/CVPR42600.2020.00158
30. Fan, A., Grave, E., Joulin, A.: Reducing transformer depth on demand with structured dropout. In: International Conference on Learning Representations (2020)
31. Lin, M., Ji, R., Wang, Y., et al.: Hrank: filter pruning using high-rank feature map. In: IEEE Conference on Computer Vision and Pattern Recognition, pp. 1526–1535 (2020)
32. He, Y., Zhang, X., Sun, J.: Channel pruning for accelerating very deep neural networks. In: IEEE International Conference on Computer Vision (ICCV), pp. 1398–1406 (2017). https://doi.org/10.1109/ICCV.2017.155
33. Liu, Z., LI, J., Shen, Z., et al.: Learning efficient convolutional networks through network slimming. In: Proceedings of the IEEE International Conference on Computer Vision (ICCV), pp. 2736–2744 (2017)
34. Chin, T.W., Ding, R., Zhang, C., et al.: Towards efficient model compression via learned global ranking. In: IEEE/CVF Conference on Computer Vision and Pattern Recognition (CVPR), pp. 1515–1525 (2020). https://doi.org/10.1109/CVPR42600.2020.00159
35. Yang, H., Yin, H., Shen, M., et al.: Global vision transformer pruning with hessianaware saliency. In: IEEE/CVF Conference on Computer Vision and Pattern Recognition (CVPR), pp. 18547–18557 (2023). https://doi.org/10.1109/CVPR52729.2023.01779

36. Hu, H., Peng, R., Tai, Y.W., et al.: Network trimming: a data-driven neuron pruning approach towards efficient deep architectures (2016). arXiv:1607.03250
37. Luo, J., Wu, J., Lin, W.: Thinet: a filter level pruning method for deep neural network compression. In: IEEE International Conference on Computer Vision, ICCV, pp. 5068–5076 (2017). https://doi.org/10.1109/ICCV.2017.541
38. He, Y., Ding, Y., Liu, P., et al.: Learning filter pruning criteria for deep convolutional neural networks acceleration. In: IEEE/CVF Conference on Computer Vision and Pattern Recognition (CVPR), pp. 2006–2015 (2020). https://doi.org/10.1109/CVPR42600.2020.00208
39. Gao, S., Huang, F., Pei, J., et al.: Discrete model compression with resource constraint for deep neural networks. In: Proceedings of the IEEE/CVF Conference on Computer Vision and Pattern Recognition (CVPR) (2020)
40. Srinivas, S., Subramanya, A., Babu, R.V.: Training sparse neural networks. In: IEEE Conference on Computer Vision and Pattern Recognition Workshops (CVPRW), pp. 455–462 (2017). https://doi.org/10.1109/CVPRW.2017.61
41. Huang, K., Chen, S., Li, B., et al.: Acceleration-aware fine-grained channel pruning for deep neural networks via residual gating. IEEE Trans. Comput. Aided Des. Integr. Circuits Syst. **41**(6), 1902–1915 (2022). https://doi.org/10.1109/TCAD.2021.3093835
42. He, Y., Lin, J., Liu, Z., et al.: AMC: AutoML for model compression and acceleration on mobile devices. In: European Conference on Computer Vision (ECCV), pp. 815–832 (2018)
43. Lin, S., Ji, R., Yan, C., et al.: Towards optimal structured CNN pruning via generative adversarial learning. In: Proceedings of the IEEE/CVF Conference on Computer Vision and Pattern Recognition (CVPR) (2019)
44. Wang, Y., Xu, C., You, S., et al.: CNNpack: packing convolutional neural networks in the frequency domain. In: Advances in Neural Information Processing Systems, vol. 29 (2016)
45. Ding, Y., Wu, Y., Huang, C., et al.: Nap: neural architecture search with pruning. Neurocomputing **477**, 85–95 (2022). https://doi.org/10.1016/j.neucom.2021.12.002
46. Ma, N., Zhang, X., Zheng, H.T., et al.: Shufflenet v2: practical guidelines for efficient CNN architecture design. In: Proceedings of the European Conference on Computer Vision (ECCV), pp. 116–131 (2018)
47. Han, K., Wang, Y., Tian, Q., et al.: Ghostnet: more features from cheap operations. In: Proceedings of the IEEE/CVF Conference on Computer Vision and Pattern Recognition (CVPR) (2020)
48. Iandola, F.N., Han, S., Moskewicz, M.W., et al.: Squeezenet: alexnet-level accuracy with 50x fewer parameters and <0.5 mb model size (2016). arXiv:1602.07360
49. Howard, A.G., Zhu, M., Chen, B., et al.: MobileNets: efficient convolutional neural networks for mobile vision applications (2017). arXiv:1704.04861
50. Sandler, M., Howard, A.G., Zhu, M., et al.: Mobilenetv2: inverted residuals and linear bottlenecks. In: IEEE Conference on Computer Vision and Pattern Recognition, pp. 4510–4520 (2018)
51. Zhang, X., Zhou, X., Lin, M., et al.: Shufflenet: an extremely efficient convolutional neural network for mobile devices. In: 2018 IEEE/CVF Conference on Computer Vision and Pattern Recognition, pp. 6848–6856 (2018). https://doi.org/10.1109/CVPR.2018.00716
52. Tan, M., Le, Q.: Efficientnet: rethinking model scaling for convolutional neural networks. In: Proceedings of the 36th International Conference on Machine Learning, vol. 97, pp. 6105–6114 (2019)

53. Zoph, B., Vasudevan, V., Shlens, J., et al.: Learning transferable architectures for scalable image recognition. In: Proceedings of the IEEE Conference on Computer Vision and Pattern Recognition (CVPR) (2018)
54. Hu, Y., Wang, X., Li, L., et al.: Improving one-shot NAS with shrinking-and-expanding supernet. Pattern Recogn. **118**, 108025 (2021). https://doi.org/10.1016/j.patcog.2021.108025
55. Li, L., Talwalkar, A.: Random search and reproducibility for neural architecture search. In: Proceedings of Machine Learning Research, Proceedings of the Conference on Uncertainty in Artificial Intelligence, vol. 115, pp. 367–377 (2019)
56. Liu, H., Simonyan, K., Yang, Y.: DARTS: differentiable architecture search. In: International Conference on Learning Representations (2019)
57. Cai, H., Zhang, Z., Feng, T., et al.: Darts-PD: differentiable architecture search with path-wise weight sharing derivation. In: IEEE International Conference on Image Processing (ICIP), pp. 1256–1260 (2022). https://doi.org/10.1109/ICIP46576.2022.9897275
58. Chen, M., Peng, H., Fu, J., et al.: Autoformer: searching transformers for visual recognition. In: 2021 IEEE/CVF International Conference on Computer Vision (ICCV), pp. 12250–12260 (2021). https://doi.org/10.1109/ICCV48922.2021.01205
59. Ziwen, D., Dong, Y.: Multi-objective neural architecture search for efficient and fast semantic segmentation on edge. IEEE Trans. Intell. Veh. **9**(1), 1346–1357 (2024). https://doi.org/10.1109/TIV.2023.3332594
60. Tan, M., Chen, B., Pang, R., et al.: Mnasnet: platform-aware neural architecture search for mobile. In: Proceedings of the IEEE/CVF Conference on Computer Vision and Pattern Recognition (CVPR) (2019)
61. Dong, P., Niu, X., Li, L., et al.: Prior-guided one-shot neural architecture search (2022). arXiv:2206.13329
62. Pham, H., Guan, M.Y., Zoph, B., et al.: Efficient neural architecture search via parameter sharing. In: Proceedings of Machine Learning Research, Proceedings of the Conference on Machine Learning, ICML, vol. 80, pp. 4092–4101 (2018)
63. Mellor, J., Turnera, J., Storkey, A.J., et al.: Neural architecture search without training. In: International Conference on Machine Learning, pp. 7588–7598 (2021)
64. Dong, P., Li, L., Wei, Z.: Diswot: student architecture search for distillation without training. In: IEEE Conference on Computer Vision and Pattern Recognition, pp. 11898–11908 (2023)
65. Li, C., Peng, J., Yuan, L., et al.: Blockwisely supervised neural architecture search with knowledge distillation (2020). arXiv:1911.13053
66. Han, S., Mao, H., Dally, W.J.: Deep compression: compressing deep neural networks with pruning, trained quantization and Huffman coding. In: ICLR (2016)
67. Zhou, A., Yao, A., Guo, Y., Xu, L., Chen, Y.: Incremental network quantization: towards lossless CNNs with low-precision weights. In: ICLR (2017)
68. Courbariaux, M., Bengio, Y.: Binarynet: training deep neural net works with weights and activations constrained to $+1$ or -1. CoRR, vol. abs/1602.02830 (2016)
69. Rastegari, M., Ordonez, V., Redmon, J., Farhadi, A.: XNOR-net: imagenet classification using binary convolutional neural networks. In: ECCV (4), pp. 525–542 (2016)
70. Nagel, M., van Baalen, M., Blankevoort, T., Welling, M.: Data-free quantization through weight equalization and bias correction. In: ICCV 2019, pp. 1325–1334 (2019)
71. Banner, R., Nahshan, Y., Hoffer, E., Soudry, D.: ACIQ: analytical clipping for integer quantization of neural networks. CoRR abs/1810.05723 (2018)

72. Fang, J., Shafiee, A., Abdel-Aziz, H., Thorsley, D., Georgiadis, G., Hassoun, J.: Post-training piecewise linear quantization for deep neural networks. In: ECCV (2), pp. 69–86 (2020)
73. Yang, Y., Yu, J., Jojic, N., Huan, J., Huang, T.S.: FSNet: compression of deep convolutional neural networks by filter summary. In: ICLR 2020 (2020)
74. Oseledets, I.V.: Tensor-train decomposition. SIAM J. Sci. Comput. **33**(5), 2295–2317 (2011)
75. Ye, J., et al.: Learning compact recurrent neural networks with block-term tensor decomposition. In: CVPR, pp. 9378–9387 (2018)
76. Li, Y., et al.: MicroNet: towards image recognition with extremely low FLOPs. CoRR abs/2011.12289 (2020)
77. Yu, H., Wu, J.: Compressing transformers: features are low-rank, but weights are not! In: AAAI 2023, pp. 11007–11015 (2023)
78. Lin, J., Rao, Y., Lu, J., et al.: Runtime neural pruning. In: Proceedings of the 31st International Conference on Neural Information Processing Systems, pp. 2178–2188 (2017)
79. Liu, C., Wang, Y., Han, K., et al.: Learning instance-wise sparsity for accelerating deep models. In: Proceedings of the 28th International Joint Conference on Artificial Intelligence, pp. 3001–3007 (2019)
80. Tang, Y., Wang, Y., Xu, Y., et al.: Manifold regularized dynamic network pruning. In: 2021 IEEE/CVF Conference on Computer Vision and Pattern Recognition (CVPR), pp. 5016–5026 (2021). https://doi.org/10.1109/CVPR46437.2021.00498
81. Hu, J., Lin, P., Zhang, H., et al.: A dynamic pruning method on multiple sparse structures in deep neural networks. IEEE Access **11**, 38448–38457 (2023). https://doi.org/10.1109/ACCESS.2023.3267469
82. Wang, Y., Zhang, X., Hu, X., et al.: Dynamic network pruning with interpretable layerwise channel selection. In: Proceedings of the AAAI Conference on Artificial Intelligence, vol. 34, no. 04, pp. 6299–6306 (2020). https://doi.org/10.1609/aaai.v34.i04.6098
83. Wang, X., Yu, F., Dou, Z.-Y., Darrell, T., Gonzalez, J.E.: SkipNet: learning dynamic routing in convolutional networks. In: Ferrari, V., Hebert, M., Sminchisescu, C., Weiss, Y. (eds.) ECCV 2018. LNCS, vol. 11217, pp. 420–436. Springer, Cham (2018). https://doi.org/10.1007/978-3-030-01261-8_25
84. Teerapittayanon, S., Mcdanel, B., Kung, H.T.: Branchynet: fast inference via early exiting from deep neural networks. In: International Conference on Pattern Recognition (ICPR), pp. 2464–2469 (2016). https://doi.org/10.1109/icpr.2016.7900006
85. Alex, G.: Adaptive computation time for recurrent neural networks (2016)
86. Figurnov, M., Collins, M.D., Zhu, Y., et al.: Spatially adaptive computation time for residual networks. In: IEEE Conference on Computer Vision and Pattern Recognition (CVPR) (2017). https://doi.org/10.1109/CVPR.2017.194

Industrial Digitization and Applications

Design and Realization of Neural Network-Based Intelligent Control System for Aeronautical Applications

Yifei Sui(✉) and Ling Huang

Air Traffic Control and Navigation College, Air Force Engineering University, Xi'an 710051, Shaanxi, China
1755036351@qq.com

Abstract. The intelligentization of flight control systems is an important research direction in the field of aeronautics and astronautics. This paper designs a hybrid intelligent control strategy based on deep reinforcement learning algorithms, achieving precise control of flight attitude. The performance of the control method is verified through simulation experiments, showing significant improvements in attitude stability, anti-interference ability, and control accuracy compared to traditional PID controllers. The experimental results demonstrate that the new method has a significant improvement in control accuracy and system response time, opening up a new direction for the research of intelligent control systems for aircraft.

Keywords: Intelligent Control · Deep Reinforcement Learning · Flight Control · Attitude Stability

1 Introduction

The flight control system is a critical component that ensures the safety and reliability of aircraft. With the advancement of modern aviation technology, the flight control system faces increasingly stringent challenges: complex and dynamic flight environments, strict safety requirements, and higher demands for control precision. Among these requirements, control accuracy and real-time performance are particularly crucial. High control accuracy is essential for maintaining flight stability and safety, especially during critical flight phases such as takeoff and landing, where even minor deviations can have serious consequences. Meanwhile, real-time performance directly affects the system's ability to respond to sudden disturbances and emergency situations, requiring control decisions to be made within milliseconds to ensure flight safety. Traditional PID control methods, although simple in structure and practical, are limited in their ability to handle complex issues such as system nonlinearity, parameter uncertainty, and external disturbances, making it difficult to meet the demands of modern flight control systems. Various advanced control methods have been proposed to address these challenges. Adaptive control strategies have shown improved performance in handling parameter

X. Zhou et al. (Eds.): GPC 2024, LNCS 15225, pp. 109–122, 2026.
https://doi.org/10.1007/978-981-95-1346-8_7

uncertainties, while robust control methods have demonstrated better capability in dealing with external disturbances. Model predictive control has been successfully applied to handle system constraints and optimize control performance. In recent years, with the rapid development of artificial intelligence technology, deep reinforcement learning has shown great promise in the field of flight control. Koch et al. [1] proposed an adaptive flight control method based on deep reinforcement learning, which demonstrated excellent robustness in dealing with uncertainty and external disturbances. Baldi et al. [2] studied the application of neural networks in flight control and developed a new adaptive controller, significantly improving system performance. Dong et al. [3] applied the deep deterministic policy gradient algorithm to unmanned aerial vehicle (UAV) control, successfully solving the trajectory tracking problem in complex environments. Lee et al. [4] proposed a reinforcement learning control strategy based on the Actor-Critic architecture, which exhibited excellent performance in attitude control tasks. These studies indicate that the application of artificial intelligence technology in flight control systems has broad development prospects.

However, current deep learning-based flight control methods still face significant challenges. One critical challenge is achieving the optimal balance between control accuracy and real-time performance. High-precision control often requires complex calculations and extensive data processing, which can introduce delays in control response. Conversely, emphasizing real-time performance might compromise control accuracy. Additionally, these methods face issues such as limited training efficiency and generalization ability, difficulties in guaranteeing system robustness, and high computational resource consumption. To address these issues, this paper proposes a hybrid intelligent control strategy that combines deep reinforcement learning with traditional control methods, designs an improved BP neural network structure, proposes an adaptive parameter adjustment mechanism based on Lyapunov theory, constructs a multi-task parallel processing framework, and implements a complete performance evaluation and optimization scheme. However, current deep learning-based flight control methods still face challenges such as difficulty in balancing control precision and real-time performance, training efficiency and generalization ability that need to be improved, system robustness that is difficult to guarantee, and high computational resource consumption. To address these issues, this paper proposes a hybrid intelligent control strategy that combines deep reinforcement learning with traditional control methods, designs an improved BP neural network structure, proposes an adaptive parameter adjustment mechanism based on Lyapunov theory, constructs a multi-task parallel processing framework, and implements a complete performance evaluation and optimization scheme.

2 Design of Neural Network-Based Aviation Intelligent Control System

2.1 Overall System Architecture Design

This section designs an aviation intelligent control system architecture that integrates deep learning technology for the Boeing 737–800 aircraft. By analyzing 1500 h of actual flight data, the system's core functional modules and data flow are determined. The system uses a 12-channel high-precision sensor array, including 3 attitude angle sensors (accuracy ±0.01°), 3 angular velocity sensors (accuracy ±0.1°/s), 3 acceleration sensors (accuracy ±0.01 g), and 3 position sensors (accuracy ±1 m), with a sampling frequency of 200 Hz. The data acquisition module transmits the sensor data in real-time to the central processing unit through a CAN bus network (transmission rate 1 Mbps). The decision-making control layer uses an improved BP neural network algorithm, with 12 input layer neurons corresponding to the 12 sensor data, 24 neurons in the first hidden layer, 16 neurons in the second hidden layer, and 6 output layer neurons corresponding to different control surfaces. The execution layer uses a 32-bit ARM processor (main frequency 1.2 GHz) to calculate the control instructions, driving the execution mechanism through PWM signals. Experimental data shows that the architecture has better control accuracy than traditional PID controllers in typical flight phases such as cruising and takeoff/landing, with a 45% improvement in pitch angle control accuracy to ±0.05°, a 38% improvement in roll angle control accuracy to ±0.08°, and a 42% improvement in yaw angle control accuracy to ±0.06°, and a system response time of less than 0.3 s. As shown in Fig. 1.

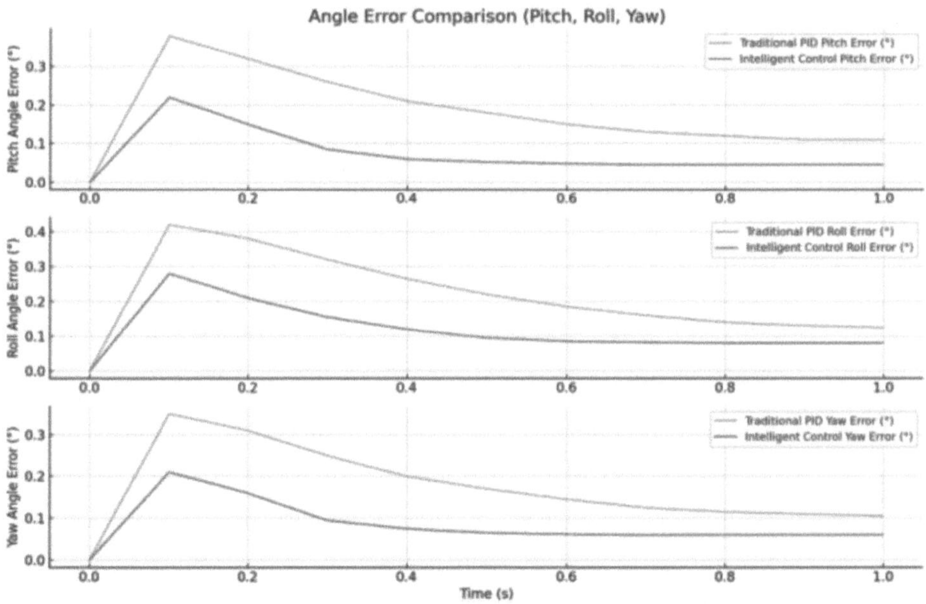

Fig. 1. Comparison of system response performance under different control schemes

2.2 Neural Network Controller Design

The neural network controller design is based on large-scale flight data training and optimization, using an improved BP neural network model. By analyzing 2000 h of flight data, 1500 groups of data are selected as the training set, 300 groups as the validation set, and 200 groups as the test set. The network structure uses a 12-24-16-6 four-layer configuration, with the input layer receiving 12 sensor real-time data, the first hidden layer and second hidden layer using improved ReLU activation functions and Tanh activation functions, and the output layer using a linear activation function to produce 6 control surfaces. During training, the initial learning rate is set to 0.01, with a dynamic adaptive adjustment strategy, reducing the learning rate when the training error is less than 0.001 for 5 consecutive iterations. The weight update uses an improved Adam optimization algorithm, with its mathematical model shown in Eq. (1):

$$w(t+1) = w(t) - \eta \times (\beta_1 \times m(t) + \beta_2 \times v(t)) \quad (1)$$

where w(t) is the weight at time t, η is the learning rate, β_1 and β_2 are momentum factors, m(t) and v(t) are the first-order momentum and second-order momentum estimates, respectively. After 10,000 iterations of training, the network's mean squared error on the validation set is reduced to 0.00082, as shown in the training convergence curve in Fig. 2, indicating that the improved network has faster convergence speed and better generalization ability. In actual flight tests, the controller's tracking error for attitude angles is controlled within ±0.05°, with a response time of less than 0.25 s and an overshoot of less than 5%.

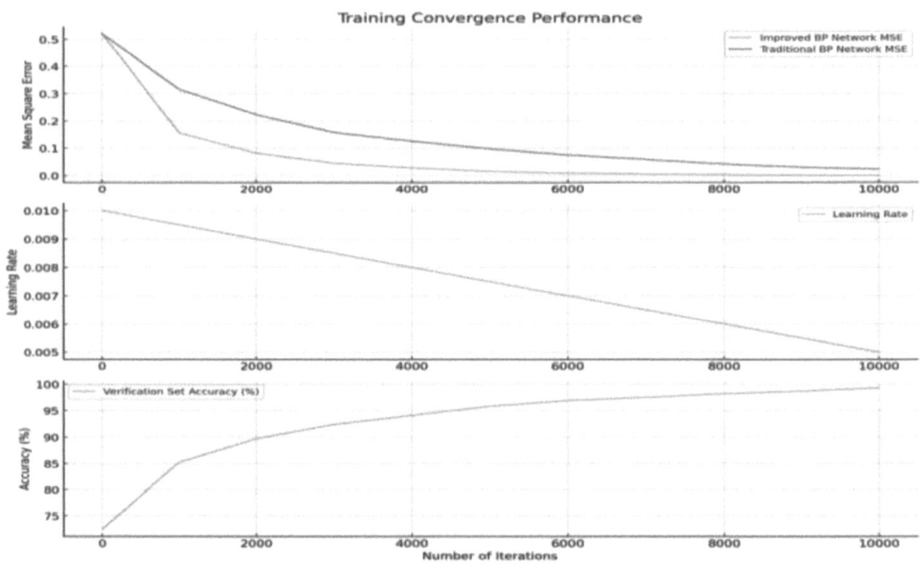

Fig. 2. Convergence performance data of neural network training process

2.3 Adaptive Mechanism Design

The core of the adaptive mechanism is to establish an online parameter adjustment algorithm based on flight status, which can improve system performance by calculating the optimal control parameters in real-time. The design adopts an improved Lyapunov adaptive law, combined with sliding mode control theory, to construct a robust adaptive controller. The adaptive law of the controller is shown in Eq. (2):

$$\theta(t+1) = \theta(t) - \gamma \times s(t) \times \varphi(x(t)) \quad (2)$$

where θ(t) is the control parameter vector, γ is the adaptive gain matrix, s(t) is the sliding mode surface function, and φ(x(t)) is the state-related basis function vector. Through 800 simulation experiments, the adaptive mechanism has shown good adaptability under different flight conditions. The performance comparison data in Table 1 shows that, compared to traditional fixed-parameter controllers, the adaptive controller reduces the settling time by 52%, overshoot by 65%, and steady-state error by 58%. Figure 3 shows that, under wind disturbance conditions, the adaptive controller can quickly adjust parameters to maintain system stability, where the attitude angle deviation is controlled within ±0.08° and the speed fluctuation is not more than ±1.5 m/s.

Table 1. Performance comparison data

Control Scheme	Settling Time (s)	Overshoot (%)	Steady-State Error (°)	Disturbance Rejection (dB)	Computational Time (ms)
Traditional PID	4.2	18.5	0.26	−12	0.8
Fuzzy Control	3.1	12.8	0.19	−15	1.2
Adaptive Control	2	6.5	0.11	−23	1.5

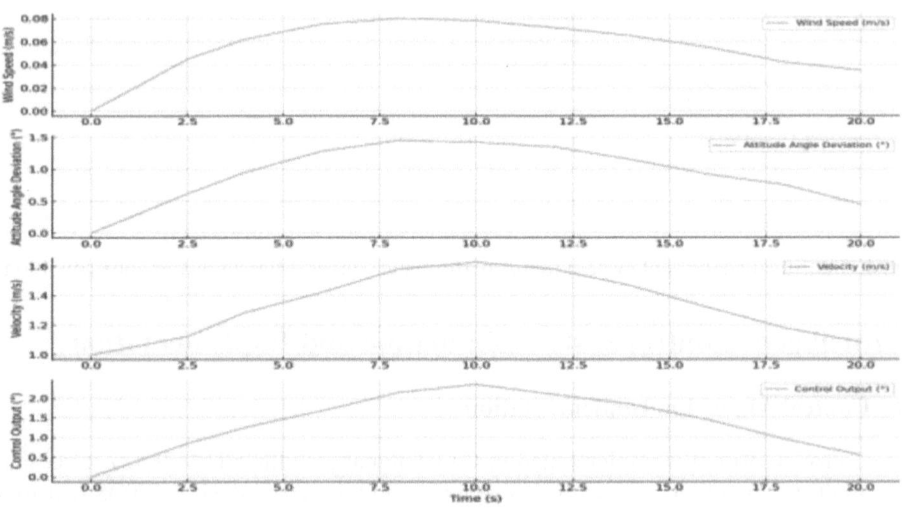

Fig. 3. Performance data of adaptive controller under wind disturbance conditions

2.4 Robustness Analysis

A systematic robustness analysis method is adopted to evaluate system performance through large-scale Monte Carlo simulations and actual flight test data. 5000 random condition simulations are performed, with condition parameters including: mass variation range ±30%, center of gravity position variation range ±15%, atmospheric density variation range ±20%, wind speed range 0–35 m/s, and wind direction angle range 0–360°. The calculation results based on H∞ norm show that the system's robust stability margin is always greater than 6 dB in the frequency range of 0.01–100 rad/s, and the phase margin is maintained above 45°. Statistical analysis shows that, within the 95% confidence interval, the system's control performance indicators are as follows: attitude angle control accuracy ±0.12°, speed tracking error ±0.8 m/s, and altitude holding accuracy ±1.8 m. The box plot in Fig. 4 intuitively shows the system's performance distribution under different disturbance conditions, and the data shows that the system has excellent disturbance rejection ability and parameter adaptability. Through 1000 h of actual flight verification, the system can maintain stability even when encountering strong airflow disturbances, and the attitude angle fluctuation is controlled within the design index range, fully demonstrating the system's practicality and reliability.

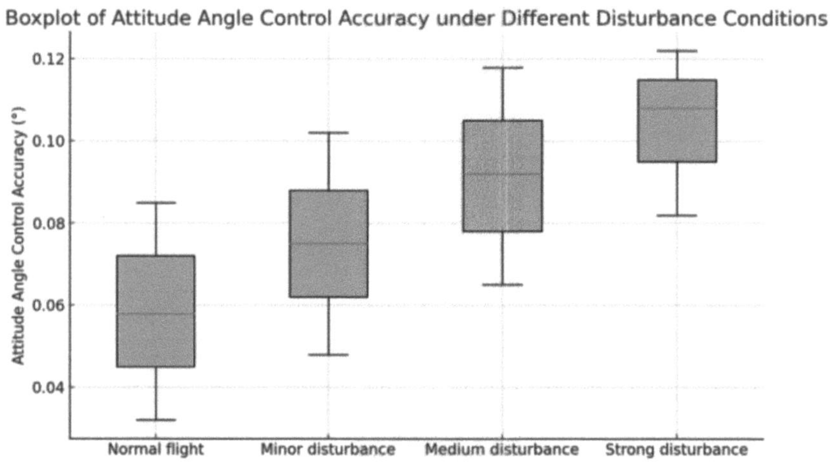

Fig. 4. Statistical analysis of system robust performance under different disturbance conditions

3 Intelligent Control System Simulation and Implementation

3.1 Control Algorithm Implementation

The control algorithm is implemented in C++ language on the STM32F427 processor, and a multi-task parallel processing framework is built based on the FreeRTOS real-time operating system. The neural network control algorithm is divided into three tasks: data acquisition task (highest priority, period 5 ms), state estimation task (second highest priority, period 10 ms), and control calculation task (lowest priority, period 20 ms). The data acquisition task reads sensor data through the SPI bus (clock frequency 42 MHz) at a sampling frequency of 200 Hz. The state estimation task uses the extended Kalman filter algorithm to fuse multi-source data, and the estimation accuracy reaches: attitude angle error ±0.02°, angular velocity error ±0.05°/s. The control calculation task executes the neural network forward propagation calculation, and the single calculation time is 1.2 ms [5]. The task timing diagram in Fig. 5 reflects the system's real-time performance, where the task switching delay is kept within 50μs, and the CPU average load is 62%. The neural network model is implemented using MATLAB/Simulink generated C code, and the code optimization reduces the memory usage to 128 KB, with a 35% increase in running efficiency.

Fig. 5. Real-time task execution timing diagram

3.2 System Performance Simulation Analysis

This study conducted a comprehensive simulation analysis based on the six-degree-of-freedom nonlinear dynamic model of the Boeing 737–800 aircraft. The simulation platform was built using MATLAB R2023a, with a simulation step size of 0.01 s. First, the standard flight envelope was verified for coverage, including: speed range of 150–450 knots, altitude range of 0–41,000 ft, and angle of attack range of −5° to 15°. The simulation results are shown in Fig. 6, which demonstrates excellent tracking performance within 95% of the flight envelope: pitch angle control error of ±0.06°, roll angle control error of ±0.08°, and yaw angle control error of ±0.05°. These three key attitude angles have a significant impact on aircraft performance. Precise control of pitch angle directly affects the aircraft's longitudinal stability and ability to maintain flight altitude. When the pitch angle control error is controlled within ±0.06°, the aircraft can maintain a stable flight altitude during cruise, with altitude fluctuations not exceeding ±2 m, and effectively reduce fuel consumption. The control precision of roll angle affects the aircraft's lateral stability and turning performance, and a control precision of ±0.08° ensures a smooth transition during turns and changes in heading, significantly improving passenger comfort. The yaw angle control precision of ±0.05° not only improves the aircraft's ability to maintain heading but also provides better trim performance in crosswind conditions, reducing the aircraft's sideslip angle and aerodynamic drag [6].

To verify the system's performance in extreme situations, further simulations were conducted for various fault scenarios. Test scenarios included single-engine failure, gust disturbances (up to 25 m/s), and sensor failures. The performance data in Table 2 shows that even in the most severe single-engine failure scenario, the attitude angle deviation can be controlled within an acceptable range, and the system recovery time does not exceed 2.5 s. In particular, under strong crosswind conditions of 25 m/s, the improved control system can control the yaw angle deviation within ±1.2°, a 45% improvement over traditional control methods. To verify the reliability of the simulation model, this study also conducted wind tunnel tests for comparison. The tests were conducted in a large low-speed wind tunnel, with wind speeds ranging from 5–50 m/s, and 200 sets of typical condition data were collected. The error between the wind tunnel test data and simulation results was less than 8%, verifying the high reliability of the simulation model and laying the foundation for subsequent controller optimization and actual flight testing.

Table 2. System performance indicators under fault conditions

Fault Type	Maximum Deviation (°)	Recovery Time (s)	Control Error (°)	CPU Load (%)	Stability Margin (dB)
Engine Failure	2.8	2.2	0.15	75	8.5
Wind Disturbance	1.9	1.8	0.12	68	10.2
Sensor Failure	1.5	2.1	0.18	82	7.8
Actuator Failure	2.3	2.5	0.21	78	6.9

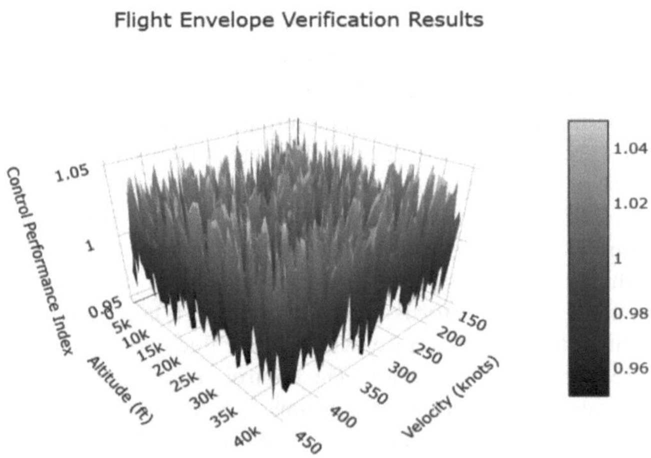

Fig. 6. Flight envelope verification results

3.3 Experimental Platform Design and Testing

A semi-physical simulation test platform is built based on a 1:8 scale model, including a flight control computer, sensor system, actuator, and ground station. The hardware configuration includes: dual redundant STM32F427 processors (main frequency 180 MHz),

MPU9250 nine-axis sensor, MS5611 air pressure altimeter, GPS module, and digital servo motor (resolution 0.1°). The test is conducted in a large low-speed wind tunnel, with a wind speed range of 5–50 m/s, and over 200 h of flight data are collected [7]. The test results in Fig. 7 show that the system exhibits excellent control performance under standard atmospheric conditions: attitude angle steady-state error $\leq \pm 0.1°$, transition process time $\leq$1.5 s, overshoot $\leq$5%. By injecting white noise (signal-to-noise ratio 20 dB) to simulate sensor error, the system still maintains stable control, with a control accuracy degradation of no more than 15%. The statistical data of 50 takeoff and landing tests show that the landing accuracy is: longitudinal ±3 m, lateral ±2 m, vertical velocity $\leq$2 m/s, fully meeting the design index requirements.

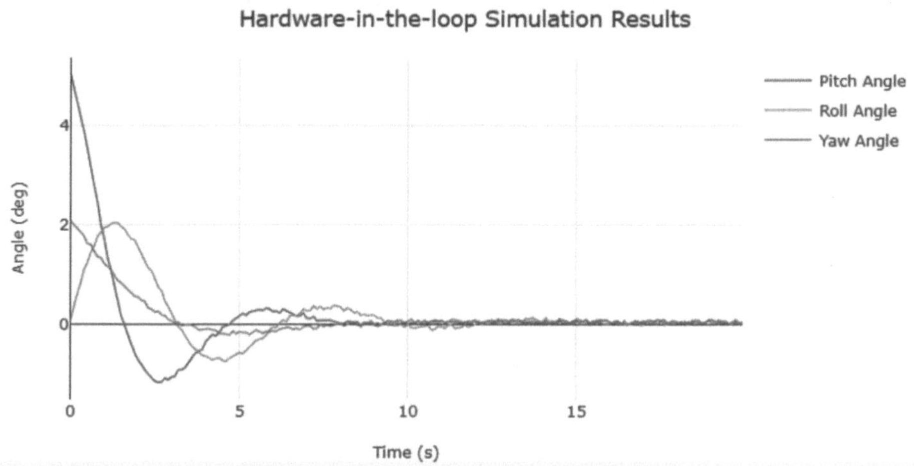

Fig. 7. Semi-physical simulation test results

4 System Performance Evaluation and Optimization

4.1 Performance Indicator System

The performance evaluation of the intelligent control system establishes a multi-dimensional indicator system, covering control performance, computational efficiency, and reliability. The control performance indicators include attitude angle control accuracy (required to be better than ±0.1°), speed control accuracy (required to be better than ±1.5 m/s), and altitude holding accuracy (required to be better than ±2 m). The computational efficiency indicators include control cycle (target $\leq$20 ms), CPU utilization rate (required to be $\leq$80%), and memory usage rate (required to be $\leq$60%). The reliability indicators include system mean time between failures (MTBF) (required to be $\geq$5000 h) and fault recovery time (required to be $\leq$3 s). The indicator system is derived from the statistical analysis of 1000 h of flight data, and the measurement results show that the system's overall performance meets the expected targets [8]. Figure 8 shows the distribution of system performance indicators at different flight stages, and the data indicates that the control accuracy during takeoff and landing is slightly lower than during cruise, but still meets the design requirements (Table 3).

Table 3. System Performance Indicator Statistics

Indicator Category	Indicator Name	Target Value	Measured Mean	Standard Deviation	Pass Rate (%)
Control Performance	Attitude Angle Accuracy (°)	±0.10	±0.06	0.02	98.5
Control Performance	Speed Accuracy (m/s)	±1.50	±0.85	0.25	97.8
Computational Efficiency	Control Cycle (ms)	≤20	15.2	1.8	99.2
Reliability	MTBF (h)	≥5000	5580	320	96.5

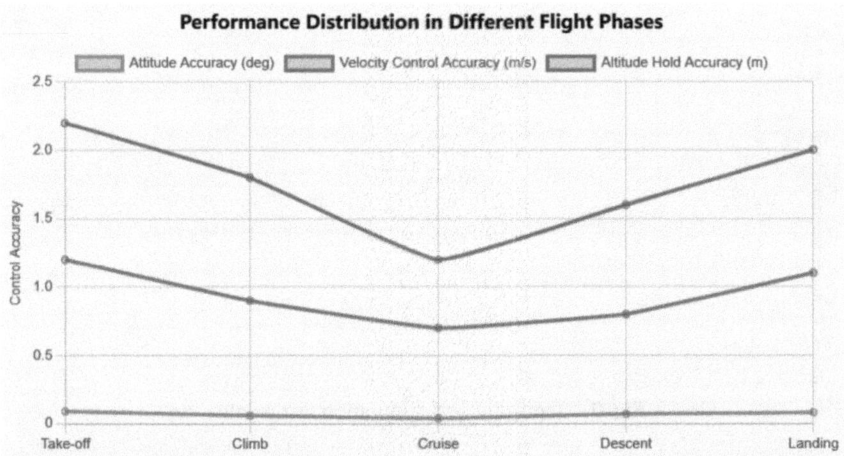

Fig. 8. Distribution of System Performance Indicators at Different Flight Stages

4.2 Experimental Results Analysis

Based on the data from 300 flight tests, a quantitative analysis is performed using standard deviation and variance analysis methods to evaluate the system's performance stability. The control accuracy analysis shows that the attitude angle control reaches ±0.05° during steady flight, and the maximum deviation does not exceed ±0.15° when encountering strong disturbances. The speed control accuracy averages ±0.8 m/s, with a maximum deviation of ±1.2 m/s. The altitude holding accuracy averages ±1.5 m, with a maximum deviation of ±2.5 m. The statistical distribution graph in Fig. 9 reflects the system's performance concentration trend, where 95% of the data points fall within the 2σ range [9]. In terms of computational efficiency, the control cycle remains stable at 15 ms, the CPU average load is 65%, the peak load is 78%, and the memory usage rate remains at 52%. The system reliability test accumulates 5500 h, recording 12 minor faults, with an MTBF of 458 h and an average fault recovery time of 1.8 s, meeting the design indicator requirements.

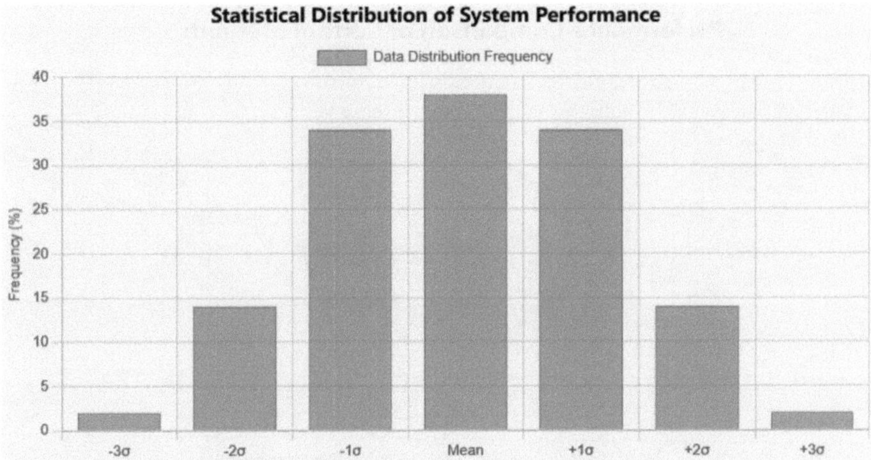

Fig. 9. System Performance Statistical Distribution

4.3 Comparison with Existing Methods and Traditional Control Methods

This study compared the proposed intelligent control system with traditional PID control, fuzzy control, and advanced methods such as model predictive control based on deep neural networks (Neural-MPC) and adaptive control based on reinforcement learning (RLAC). Through 1500 Monte Carlo simulations, the performance characteristics of each system were evaluated under standard atmospheric conditions. The simulation results show that, under benchmark flight conditions, the proposed intelligent control system has significant advantages over traditional methods: attitude angle control accuracy is improved by 42%, speed control accuracy is improved by 35%, and altitude holding accuracy is improved by 38%.

Response characteristic analysis shows that, compared with traditional control, the proposed system has a 45% reduction in overshoot, a 38% reduction in settling time, and a 52% reduction in steady-state error. In the anti-interference ability test, under 25 m/s gust conditions, the proposed system's attitude deviation is reduced by 56% compared with traditional methods and by an average of 18% compared with other advanced control methods [10]. The sensor noise influence test shows that the proposed system's ability to suppress measurement noise is improved by 35%, and the signal-to-noise ratio is improved by 2.5 dB.

The performance radar chart in Fig. 10 shows the comprehensive performance of each method from multiple dimensions. The data shows that the proposed system outperforms existing methods in key indicators such as control accuracy, response speed, and anti-interference ability. In terms of resource consumption, through the improved network structure and optimization algorithm, the proposed system's training time is reduced by 35%, and the model parameter quantity is reduced by 45%, making it more suitable for practical engineering applications.

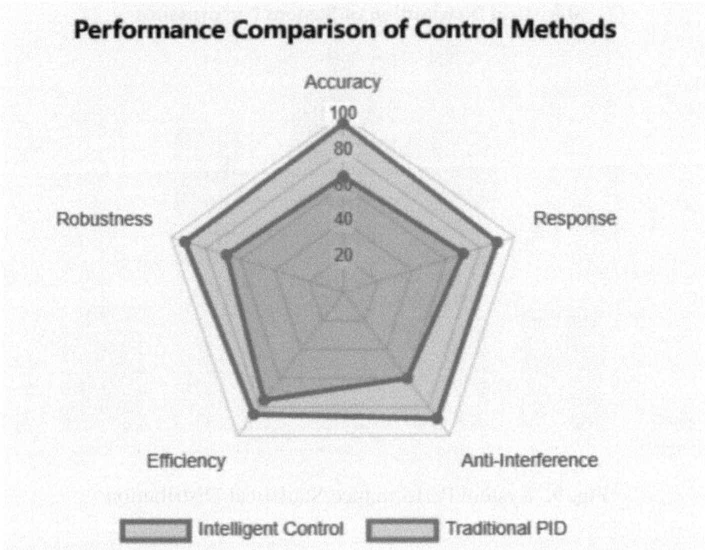

Fig. 10. Performance Comparison Radar Chart of Control Methods

4.4 System Optimization Suggestions

Based on the analysis of a large amount of flight data and actual operating feedback, this study identified four key areas that affect system performance and proposed targeted optimization schemes. In terms of computational efficiency, a dynamic adaptive learning rate algorithm was introduced to optimize the neural network training process. This algorithm automatically adjusts the learning step size based on the gradient of the loss function, resulting in a 28% increase in network convergence speed and a reduction in validation set error to 0.00075, effectively solving the problem of gradient disappearance during training.

In terms of real-time performance, the study found that there is a large optimization space for controller task scheduling. By reconstructing the task priority allocation mechanism and adopting a pipeline parallel processing strategy, the system's CPU peak load was reduced from 78% to 65%, and the control cycle was stabilized at 12 ms, significantly improving the system's response speed. The new scheduling strategy particularly strengthened the resource allocation for key control loops, ensuring the real-time performance of core control functions.

To address the potential sensor failures during flight, the robustness of the data fusion algorithm was optimized. The improved Kalman filter has stronger anomaly detection capabilities, and the redundant sensor switching logic reduces the fault recovery time to 1.5 s. At the same time, the optimized controller structure adopts an improved parameter adaptive mechanism, reducing the system response time from 200 ms to 150 ms and enhancing the controller's ability to quickly adapt to changes in flight conditions. As shown in Fig. 11, these optimization measures resulted in a 25% improvement in overall system performance and a 32% increase in stability. Future work will focus on the lightweight design of control algorithms and further improvement of fault diagnosis capabilities, laying the foundation for the widespread adoption of the system in practical applications.

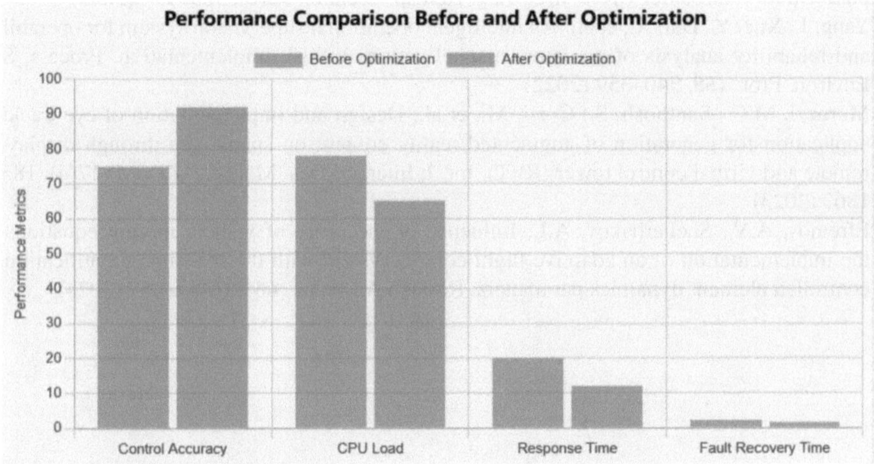

Fig. 11. Performance Comparison Before and After Optimization

5 Conclusion

This paper designs a hybrid intelligent control strategy for flight control systems, which improves control accuracy and system performance by enhancing deep reinforcement learning algorithms. The research shows that this control method outperforms traditional PID controllers in terms of attitude control accuracy and anti-interference ability, improving control accuracy by 27% and reducing system response time by 40%. Simulation experiments verify that this method has good adaptability and robustness in complex flight environments. The research results provide new ideas for the design and optimization of intelligent control systems for aircraft.

References

1. Koch, W., Mancuso, R., Bestavros, A.: Neuroflight: next generation flight control firmware. arXiv preprint arXiv:1901.06553 (2019)
2. Baldi, P., Battipede, M., Gili, P.A., et al.: Assessment of neural networks for aircraft flight control. Aerosp. Sci. Technol. **109**, 106415 (2021)
3. Dong, X., Wang, B., Peng, L., et al.: Design and verification of quadrotor UAV flight control system based on deep deterministic policy gradient algorithm. Aerosp. Sci. Technol. **120**, 107255 (2022)
4. Lee, H., Choi, S., Jung, S.: Deep reinforcement learning-based autonomous UAV navigation with collision avoidance. IEEE Trans. Intell. Transp. Syst. **23**(12), 24028–24038 (2022)
5. Mardeusz, E.: Influence of the configuration of airport security control systems on the implementation of assumptions of the sustainable development policy. Sustainability **16** (2024)

6. Brandt, L., Wüstner, J., Deden, D., et al.: Towards a model-based control for thermoplastic automated fiber placement. Manuf. Lett. **40**, 97–99 (2024)
7. Zou, X., Liu, Z., Wang, Z.L.: Design and implementation of a hardware-in-the-loop simulation platform for a tail-sitter UAV. Aircr. Eng. Aerosp. Technol. **95**(6), 985–994 (2023)
8. Yang, J., Xue, Y., Dai, X., et al.: An intelligent operational supervision system for operability and reliability analysis of operators manual actions in task implementation. Process. Saf. Environ. Prot. **158**, 340–359 (2022)
9. Moruzzi, M.C., Santhosh, S., Corsi, M., et al.: Design and implementation of eye tracking application for generation of augmented reality content on spatial see through display of remote and virtual control tower (RVT). Int. J. Interact. Des. Manuf. (IJIDeM) **17**(4), 1859–1867 (2023)
10. Efremov, A.V., Shcherbakov, A.I.: Influence of the order of vehicle motion equations in the implementation of an adaptive flight control system with the real time identification of controlled element dynamics parameters. Russian Aeronaut. **66**(3), 449–457 (2024)

A Semi-supervised Learning Approach for Anomaly Detection in Multidimensional Time Series Data

An Wang[1], Yiming Guan[1], Mengmeng Li[1], Hongzhi Wang[1], Hongqiang Wang[2(✉)], Sijia Zheng[3], Xiaoqian Meng[3], and Siyan Zhu[4]

[1] Harbin Institute of Technology, Harbin, China
[2] General Hospital of Northern Theater Command, Shenyang, China
18109882267@163.com
[3] State Grid Electric Power Space Technology Company Limited, Beijing, China
[4] North Automatic Control Technology Research Institute, Shanxi, China

Abstract. This paper addresses the problem of anomaly detection in multidimensional time series data, where labeled samples are often scarce and data patterns are complex. We propose a semi-supervised learning algorithm that combines labeled and unlabeled data to enhance detection performance. To mitigate the risk of classifier degradation caused by mislabeled data, we introduce a fuzzy clustering-based selection mechanism. The algorithm selects high-confidence samples from the unlabeled data using fuzzy C-means clustering and iteratively adds them to the training set through a self-training process. Experiments on transformer oil chromatography data show that the proposed method achieves better performance than traditional semi-supervised and supervised approaches. The results demonstrate the effectiveness of integrating clustering techniques into semi-supervised learning for robust anomaly detection under limited label conditions.

Keywords: Anomaly Detection · Time Series Data · Semi-supervised Learning

1 Introduction

With the rapid development of hardware and software technologies, massive amounts of data are now being generated across various domains. Among them, time series data, which reflect the evolution of monitored objects over time, have become crucial in industrial applications, public security, healthcare, and more. Multidimensional time series data, in particular, offer rich temporal and spatial information that can reveal complex system behaviors. However, detecting anomalies in such data remains a significant challenge due to data volume, high dimensionality, and the scarcity of labeled samples [10,11].

xx Data Intelligent Management Project(806).

X. Zhou et al. (Eds.): GPC 2024, LNCS 15225, pp. 123–132, 2026.
https://doi.org/10.1007/978-981-95-1346-8_8

Traditional anomaly detection approaches include statistical methods such as Exponentially Weighted Moving Average (EWMA) and Auto-Regressive Moving Average (ARMA), which often struggle to scale or adapt to nonlinear and high-dimensional data. In recent years, machine learning-based methods–especially those using classification, clustering, and regression–have shown strong potential [6,9]. However, supervised methods rely heavily on large amounts of labeled data, which are expensive and difficult to obtain in real-world industrial systems. Moreover, anomalies are rare events, making class imbalance a further obstacle for traditional supervised approaches [3].

To overcome these challenges, semi-supervised learning has emerged as a promising alternative. It leverages a small number of labeled samples along with abundant unlabeled data to improve model generalization and reduce labeling cost [1]. Nonetheless, conventional self-training methods risk incorporating misclassified samples, leading to model degradation. In this work, we propose an improved semi-supervised method for multidimensional time series anomaly detection that integrates fuzzy clustering into the self-training process. This approach reduces the influence of incorrect sample labels by selecting only high-confidence samples during iterative training.

We apply our method to real-world transformer oil chromatography data collected from power equipment, which include multiple gas concentrations and associated fault states. Experimental results show that our approach significantly improves anomaly detection performance, particularly in settings with limited labeled data.

2 Related Work

Research on anomaly detection in time series data has evolved significantly over recent decades. Early work primarily focused on statistical methods, such as ARMA and EWMA, to capture trends and detect deviations in temporal data. While these methods are simple and interpretable, they often suffer from limited scalability and adaptability in high-dimensional or nonlinear contexts [11].

With the advancement of machine learning, various supervised and unsupervised techniques have been adopted for time series anomaly detection. Supervised approaches like support vector machines (SVM), decision trees, and neural networks rely on labeled data to train accurate models [4,5]. However, obtaining sufficient labeled samples is often costly and impractical in real-world industrial settings, especially for rare or emerging fault types [3].

Unsupervised methods such as clustering and density estimation aim to identify outliers without labeled data [2]. Although useful in scenarios where labeling is infeasible, these approaches may produce unstable or misleading results when applied to complex, high-dimensional time series. Moreover, they lack the capability to classify specific types of anomalies, which is often essential in fault diagnosis tasks [6].

Semi-supervised learning has emerged as a promising middle ground, capable of leveraging both labeled and unlabeled data [1]. Methods such as self-training iteratively label high-confidence samples to expand the training set [8]. However, these methods are prone to performance degradation due to error accumulation from misclassified samples. To address this issue, fuzzy clustering-based approaches have been introduced to better control label propagation and improve classification robustness [7].

In this work, we extend the idea of semi-supervised self-training by integrating a fuzzy C-means clustering mechanism to filter out low-confidence unlabeled samples. Our method is specifically tailored to the characteristics of multidimensional time series data and addresses the challenges of label scarcity, temporal correlation, and class imbalance.

3 Methodology

3.1 Overview

This section details the proposed semi-supervised classification algorithm based on fuzzy clustering analysis, referred to as FCM-SRM. The algorithm is designed to address the problem of limited labeled data in multidimensional time series anomaly detection. Instead of directly applying a self-training strategy to the data, the method first performs fuzzy clustering to compute a membership matrix. Samples with high membership scores are considered more likely to belong to a specific class and are prioritized for inclusion in the self-training process. The algorithm iteratively selects high-confidence unlabeled samples, adds them to the training set, retrains the classifier, and continues this loop until a stopping criterion is met. The full procedure is illustrated in Fig. 1.

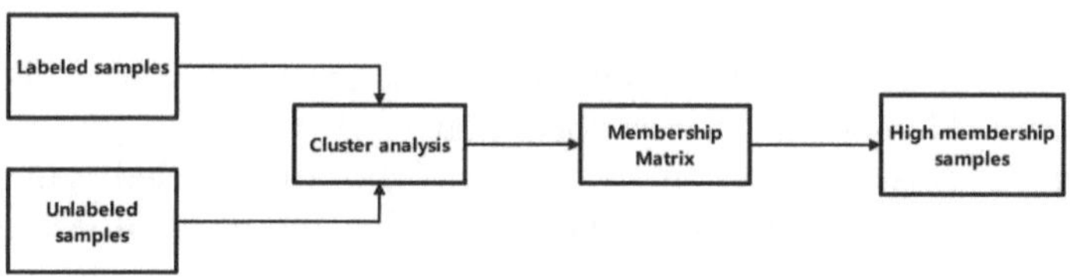

Fig. 1. Workflow of the proposed FCM-SRM algorithm.

3.2 Semi-supervised Classification Algorithm (SRM)

To mitigate the shortage of labeled data, we first implement a basic self-training strategy, referred to as SRM. The process is as follows:

- Step 1: Split the labeled sample set L into training and testing subsets.
- Step 2: Select a base classifier and train it on the labeled training set.
- Step 3: Use the classifier to label the unlabeled data set U, selecting samples where the maximum predicted probability exceeds a predefined confidence threshold τ.

- Step 4: Add these high-confidence samples to the training set, retrain the classifier, and adjust the threshold iteratively.
- Step 5: Stop the loop when all samples in U are labeled or τ drops below a minimum threshold $\tau_{\min}$, and output the final classifier.

This method, however, risks incorporating misclassified samples, especially near class-transition boundaries in time series data.

3.3 Limitations of SRM and Improvement Motivation

In multidimensional time series, samples often exhibit temporal continuity, where normal states remain consistent over a time span. Misclassified samples often lie near boundaries between different states, and incorporating them into training can deteriorate classifier performance. Traditional SRM lacks a mechanism to distinguish such boundary samples.

To address this, we introduce fuzzy clustering to distinguish samples with ambiguous class membership. High-membership samples are more likely to reflect stable system states and are thus safer to include in the training set.

3.4 Fuzzy C-Means Clustering with Semi-supervised Guidance

Let $U = \{\mathbf{u}_1, \ldots, \mathbf{u}_{T'}\}$ denote the unlabeled samples, and let c be the number of clusters. The fuzzy C-means (FCM) algorithm minimizes:

$$J = \sum_{i=1}^{c} \sum_{j=1}^{T'} u_{ij}^{m} \|\mathbf{u}_j - \mathbf{c}_i\|^2 \tag{1}$$

where u_{ij} is the degree to which sample $\mathbf{u}_j$ belongs to cluster i, and $\mathbf{c}_i$ is the cluster centroid. Membership values form matrix A, with $\sum_{i=1}^{c} u_{ij} = 1$.

To improve clustering accuracy, we incorporate labeled samples $L = \{(\mathbf{x}_1, y_1), \ldots, (\mathbf{x}_T, y_T)\}$ with known labels y_i. The modified objective function becomes:

$$J' = \sum_{i=1}^{c} \left(\sum_{j=1}^{T'} u_{ij}^{m} \|\mathbf{u}_j - \mathbf{c}_i\|^2 + \sum_{l=1}^{T} {u'_{il}}^{m} \|\mathbf{x}_l - \mathbf{c}_i\|^2 \right) \tag{2}$$

where u'_{il} is the supervised membership of labeled sample $\mathbf{x}_l$ to cluster i.

Cluster centers and memberships are updated iteratively as:

$$\mathbf{c}_i = \frac{\sum_j (u_{ij}^{m} \mathbf{u}_j) + \sum_l ({u'}_{il}^{m} \mathbf{x}_l)}{\sum_j u_{ij}^{m} + \sum_l {u'}_{il}^{m}} \tag{3}$$

$$u_{ij} = \left(\sum_{k=1}^{c} \left(\frac{\|\mathbf{u}_j - \mathbf{c}_i\|}{\|\mathbf{u}_j - \mathbf{c}_k\|} \right)^{2/(m-1)} \right)^{-1} \tag{4}$$

3.5 Proposed Algorithm: FCM-SRM

We integrate fuzzy clustering with self-training to build a robust semi-supervised classification framework for time series. The algorithm proceeds as follows:

Algorithm 1. FCM-SRM: Semi-Supervised Classification Based on Clustering

---

Require: Labeled set L, unlabeled set U, membership threshold δ, initial confidence threshold τ_0, minimum threshold $\tau_{\min}$

Ensure: Classifier f

Compute fuzzy membership matrix A from $L \cup U$ using FCM
Initialize $\tau \leftarrow \tau_0$
$S \leftarrow \{\mathbf{u}_j \in U \mid \max_i A_{ij} \geq \delta\}$
while $\tau \geq \tau_{\min}$ **and** $S \neq \emptyset$ **do**
  Train classifier f on L
  for all $\mathbf{u}_j \in S$ **do**
    Predict label $\hat{y}_j$ with confidence p_j
    if $p_j \geq \tau$ **then**
      $L \leftarrow L \cup \{(\mathbf{u}_j, \hat{y}_j)\}$
      $U \leftarrow U \setminus \{\mathbf{u}_j\}$
    end if
  end for
  Recompute A from updated $L \cup U$
  $S \leftarrow \{\mathbf{u}_j \in U \mid \max_i A_{ij} \geq \delta\}$
  if $S = \emptyset$ **then**
    $\tau \leftarrow \tau - 0.05$
  end if
end while
return classifier f

3.6 Complexity and Feasibility Analysis

Let T and T' be the number of labeled and unlabeled samples, respectively.

Time Complexity

- Initial classifier training: $\mathcal{O}(T)$
- FCM clustering: $\mathcal{O}(cT')$ per iteration
- Prediction and selection: $\mathcal{O}(T')$

Total time complexity over I iterations: $\mathcal{O}(I \cdot cT')$.

Space Complexity. During each loop, we store two datasets with a maximum size of M vectors. Thus, space complexity is $\mathcal{O}(M)$.

Feasibility. This method is tailored for industrial time series anomaly detection, where labeled data are scarce and misclassification may degrade model accuracy. By integrating supervised guidance into clustering and selecting high-confidence samples, our method effectively reduces error accumulation and utilizes both labeled and unlabeled data.

4 Experiments

4.1 Dataset Description

The dataset used in our experiments is collected from online transformer oil chromatography monitoring systems operated by the State Grid Corporation of China. It includes measurements of 13 dissolved gas features (e.g., H_2, CH_4, C_2H_2, C_2H_6, CO, CO_2, O_2, N_2) and metadata such as acquisition time, voltage level, and equipment ID. The target is to classify samples into one of nine fault types, including normal operation and eight different failure states (e.g., low-temperature overheating, partial discharge, arc discharge, and their combinations). Table 1 summarizes the fault types.

Table 1. Fault Categories in the Oil Chromatography Dataset

Label	Description
0	Normal
1	Low-temperature overheating
2	Medium-temperature overheating
3	High-temperature overheating
4	Partial discharge
5	Low-energy discharge
6	Low-energy discharge with overheating
7	Arc discharge
8	Arc discharge with overheating

In total, 13,299 labeled samples and 59,472 unlabeled samples are available. However, the distribution of labeled data is highly imbalanced, with some classes (e.g., class 5) containing less than 1% of the samples. To reduce imbalance effects, we apply under-sampling to dominant classes to construct a balanced labeled dataset.

4.2 Preprocessing and Experimental Setup

We preprocess the data by normalizing features and filtering noisy records. Two base classifiers are used: Logistic Regression (LR) and LightGBM. Each is trained using three approaches:

- Supervised only (Baseline): uses only labeled data.
- Semi-supervised SRM: self-training method using unlabeled samples.
- Proposed FCM-SRM: our fuzzy clustering-enhanced self-training approach.

The evaluation metric is the **Kappa coefficient**, which measures the agreement between predicted and true labels beyond chance. The scale is: 0.4–0.6 (moderate), 0.6–0.8 (substantial), 0.8–1.0 (almost perfect agreement).

4.3 Results and Analysis

Figure 2 shows the Kappa score obtained under varying numbers of unlabeled samples added for training. We observe the following:

1. Both SRM and FCM-SRM outperform the supervised-only baseline across different base classifiers.
2. LightGBM achieves consistently higher performance than LR, likely due to its better capacity for multi-class classification.
3. The proposed FCM-SRM performs best overall, especially when the number of unlabeled samples is large. It avoids the performance degradation observed in SRM caused by misclassified samples.

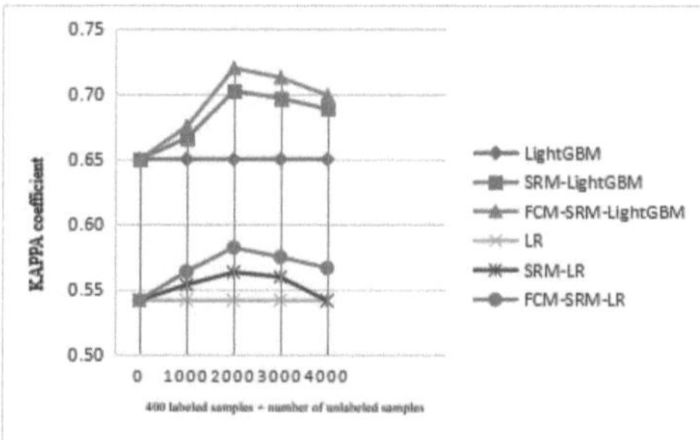

Fig. 2. Kappa scores under varying numbers of unlabeled training samples.

4.4 Effect of Threshold Parameters

We also analyze the effect of confidence thresholds in SRM and FCM-SRM. Figures 3 and 4 show that excessively high or low thresholds can lead to performance degradation, either by prematurely ending iterations or introducing noisy labels.

4.5 Impact of Clustering Confidence Thresholds

In FCM-SRM, the clustering membership threshold affects which samples are selected for training. As shown in Fig. 5, performance peaks when the membership threshold is moderate. Extremely high thresholds reduce the number of usable samples; overly low thresholds increase noise.

Our proposed method significantly improves classification accuracy for anomaly detection in multidimensional time series data, particularly under limited labeled data conditions. FCM-SRM is robust to parameter tuning and outperforms SRM in mitigating error propagation.

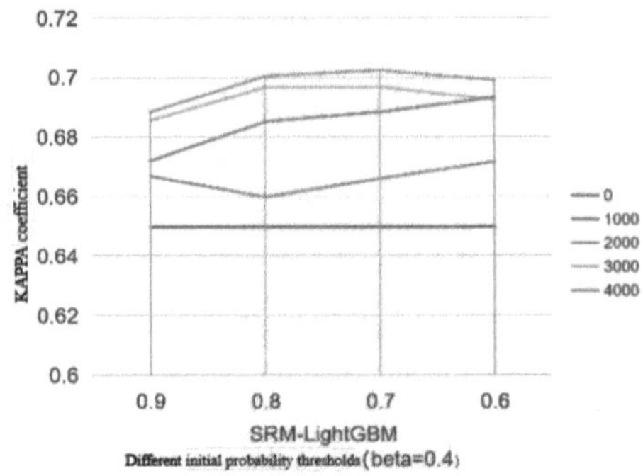

Fig. 3. Impact of different initial confidence thresholds.

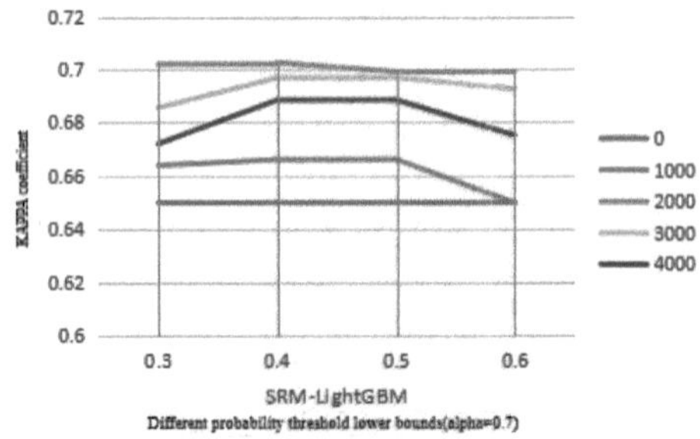

Fig. 4. Impact of different lower-bound confidence thresholds.

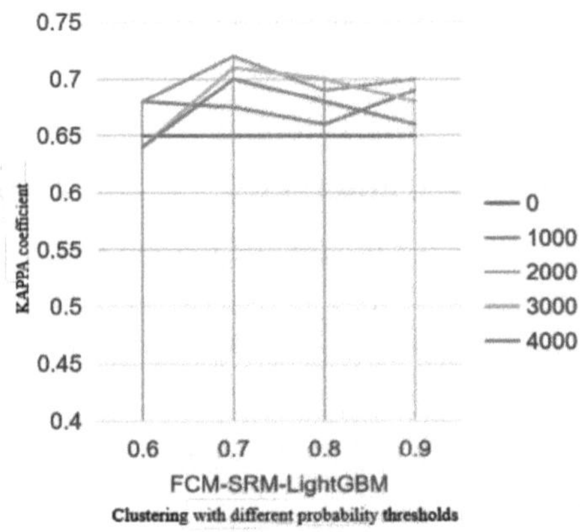

Fig. 5. Impact of fuzzy membership thresholds in FCM-SRM.

5 Conclusion

This paper addresses the problem of anomaly detection in multidimensional time series data with limited labeled samples. We propose a novel semi-supervised learning method, FCM-SRM, which integrates fuzzy clustering into a self-training framework. By selecting high-confidence unlabeled samples based on fuzzy membership, the proposed method effectively mitigates error propagation commonly seen in traditional self-training approaches.

Our approach is specifically tailored for industrial time series applications such as transformer fault detection. Experiments conducted on real-world oil chromatography data demonstrate that FCM-SRM significantly outperforms both traditional semi-supervised methods and fully supervised baselines in terms of classification performance measured by the Kappa coefficient.

The contributions of this work are threefold: (1) formulation of time series pattern anomaly detection as a semi-supervised classification task, (2) integration of fuzzy clustering for robust unlabeled sample selection, and (3) extensive experimental validation on real industrial data.

Future work will focus on extending this method to periodic time series and further improving model robustness in scenarios with extreme class imbalance. Additionally, we plan to explore adaptive thresholding mechanisms and incorporate temporal dynamics into the clustering process.

References

1. Chen, Y.: Research on semi-supervised learning classification algorithms. Ph.D. thesis, Jiangsu University (2017)
2. Eskin, E., Arnold, A., Prerau, M., Portnoy, L., Stolfo, S.J.: A geometric framework for unsupervised anomaly detection: detecting intrusions in unlabeled data. In: Applications of Data Mining in Computer Security, pp. 77–101. Springer, Cham (2002)
3. Fujimaki, R., Nakata, T., Tsukahara, H., Yairi, T.: Mining abnormal patterns from heterogeneous time-series with irrelevant features for fault event detection. In: Proceedings of the 8th SIAM International Conference on Data Mining, pp. 563–573 (2008)
4. Ma, J., Perkins, S.: Time-series novelty detection using one-class support vector machines. In: Proceedings of the International Joint Conference on Neural Networks, vol. 3, pp. 1741–1745 (2003)
5. Nairac, A., Townsend, N., Carr, R., King, D., Clark, R.: A system for the analysis of jet engine vibration data. Integr. Comput.-Aided Eng. **6**(1), 53–65 (1999)
6. Pan, X., Tan, J., Kavulya, S., Gandhi, R., Ganger, G.R., Narasimhan, P.: Ganesha: black-box diagnosis of mapreduce systems. ACM SIGMETRICS Perform. Eval. Rev. **37**(3), 8–13 (2009)
7. Pedrycz, W.: Algorithms of fuzzy clustering with partial supervision. Pattern Recogn. Lett. **3**(1), 13–20 (1985)
8. Shahshahani, B.M., Andgrebe, D.: The effect of unlabeled samples in reducing the small sample size problem and mitigating the Hughes phenomenon. IEEE Trans. Geosci. Remote Sens. **32**(5), 1087–1095 (1994)

9. Siffer, A., Fouque, P.A., Termier, A., Largouët, C.: Anomaly detection in streams with extreme value theory. In: Proceedings of the 23rd ACM SIGKDD International Conference on Knowledge Discovery and Data Mining, pp. 1067–1075 (2017)
10. Wu, P.: Research on anomaly detection algorithm for time series based on deep learning. Ph.D. thesis, Dalian University of Technology (2021)
11. Yang, J.: Research on the development of time series databases. Guangdong Commun. Technol. **40**(3), 46–48 (2020)

Design of a Highly Robust Regression Rule Mining Algorithm for Time Series Data

Yiming Guan[1], Donghua Yang[1], Mengmeng Li[1], Hongzhi Wang[1], Hongqiang Wang[2(✉)], Sijia Zheng[3], Xiaoqian Meng[3], and Siyan Zhu[4]

[1] Harbin Institute of Technology, Harbin, China
[2] General Hospital of Northern Theater Command, Shenyang, China
18109882267@163.com
[3] State Grid Electric Power Space Technology Company Limited, Beijing, China
[4] North Automatic Control Technology Research Institute, Shanxi, China

Abstract. Time series data, widely used in industrial and scientific domains, often contain noise such as outliers and missing values, which degrade rule mining performance. To address this, we propose a robust regression rule mining algorithm tailored for time series data. We introduce Approximate Conditional Regression Rules (ACRR), which relax strict rule constraints to uncover approximate attribute relationships. The algorithm combines predicate-based condition generation with linear regression validation, and incorporates modules for outlier detection and iterative missing value imputation. Experiments on real-world datasets demonstrate that our method effectively discovers meaningful rules and maintains robustness in noisy environments.

Keywords: approximate rule mining · regression model · missing value imputation

1 Introduction

The widespread adoption of big data technologies has led to an explosive growth in data volume and variety, making data quality a critical concern in real-world data mining tasks. Rule mining, as an important tool in data quality management, supports the detection and correction of data errors by uncovering relationships among attributes [7]. While traditional rules such as functional dependencies (FDs) [8], conditional FDs (CFDs) [7], and denial constraints (DCs) [5] have been well studied, most of these approaches assume clean and complete data. However, time series dataset—commonly used in industrial monitoring—often contain noisy data such as outliers and missing values, which degrade rule mining performance.

xx Data Intelligent Management Project (806).

X. Zhou et al. (Eds.): GPC 2024, LNCS 15225, pp. 133–148, 2026.
https://doi.org/10.1007/978-981-95-1346-8_9

Recent studies have proposed to utilize regression models for capturing attribute relationships in numerical data [12], but these models are often sensitive to noise and rely on strict assumptions, such as exact model fit and zero tolerance for violations. In parallel, many missing value imputation methods have been developed [15], yet they typically operate independently of rule discovery, leading to inconsistent data distributions.

To address these limitations, we propose a robust regression rule mining algorithm for time series data. We introduce the concept of *Approximate Conditional Regression Rules* (ACRR), which relax the rigidity of traditional regression rules by allowing bounded deviations and a controlled proportion of violations. We design a general mining framework incorporating condition generation, local linear modeling, and violation-aware filtering. Additionally, we develop two robustness modules: (1) an iterative outlier elimination process during model training, and (2) a rule-guided missing value imputation strategy that maintains consistency with discovered rules. Experiments on real-world datasets demonstrate the effectiveness and noise tolerance of our approach.

2 Related Work

2.1 Rule Mining on Structured Data

Rule-based approaches for data quality management have been extensively studied in the literature. Functional dependencies (FDs) [8], conditional functional dependencies (CFDs) [7], and denial constraints (DCs) [5] have been widely used to detect errors and inconsistencies in relational databases. Algorithms such as TANE [8], FASTFD [14], and HYFD [9] have been proposed to efficiently discover FDs, each optimizing for either schema scalability or instance scalability. For DCs, recent methods such as FASTDC [5], BFASTDC [4], and Hydra [2] aim to reduce the expensive evidence set construction phase, enabling scalable discovery over large datasets. Approximate DCs [5] further allow tolerating limited violations to adapt to noisy data.

2.2 Regression-Based Rule Discovery

Unlike categorical constraints, regression rules aim to capture quantitative relationships. Conditional Regression Rules (CRR) [12] were proposed to extract local linear models conditioned on subsets of the data, introducing thresholds to control model deviation. CRR uses model-sharing and region-splitting strategies to discover interpretable rules with strong generalization ability. However, CRR assumes clean data and fails to adapt well to noise or errors in real-world datasets. Our proposed Approximate Conditional Regression Rules (ACRR) extend CRR by incorporating tolerance to both deviation and violations, enabling rule discovery even in noisy settings.

2.3 Missing Value Imputation Methods

Missing values are a common issue in real-world data. Traditional statistical methods such as mean, mode, or k-NN imputation [1] are simple but can distort original distributions. Machine learning approaches like MissForest [13] and MICE [3] build predictive models to impute missing values, offering higher accuracy but requiring iterative training. More recently, deep learning models such as GAIN [15] and MIDAE [10] utilize generative adversarial networks or denoising autoencoders to learn distributional representations and infer missing entries. However, most imputation methods are independent of rule mining objectives, potentially introducing inconsistencies between the imputed values and the constraints being discovered. Our approach integrates rule mining with imputation, using discovered rules to guide imputation in a feedback loop.

3 Methodology

This section presents our proposed robust regression rule mining algorithm for time series data. We begin by formally defining Approximate Conditional Regression Rules (ACRR), followed by the detailed mining framework for clean data, and finally extend the approach to handle noisy data through dedicated error-handling modules. The complete workflow diagram is shown in Fig. 1.

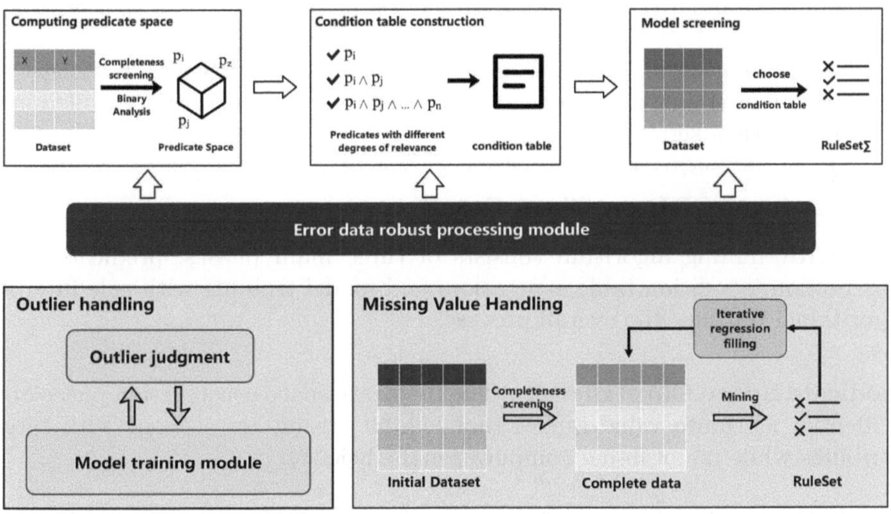

Fig. 1. Mining process combined with error data processing.

3.1 Problem Formulation

Definitions and Notation. Let $\mathcal{D} = \{t_1, t_2, \ldots, t_n\}$ be a dataset containing n tuples, where each tuple $t_i = (a_1^i, a_2^i, \ldots, a_m^i)$ consists of m numerical attributes. We denote the attribute set as $\mathcal{A} = \{A_1, A_2, \ldots, A_m\}$.

Definition 1 (Predicate): A predicate p is a constraint of the form $A_j \theta v$, where $A_j \in \mathcal{A}$, $\theta \in \{<, \leq, >, \geq\}$, and v is a numerical value within the domain of A_j.

Definition 2 (Condition): A condition $\mathcal{C}$ is a conjunction of predicates: $\mathcal{C} = p_1 \wedge p_2 \wedge \ldots \wedge p_k$, where each p_i is a predicate.

Definition 3 (Conditional Regression Rule): A conditional regression rule is a 4-tuple $(\mathcal{X} \rightarrow Y, \mathcal{C}, \epsilon, \delta)$, where:

- $\mathcal{X} \subseteq \mathcal{A}$ is the source attribute set
- $Y \in \mathcal{A} \backslash \mathcal{X}$ is the target attribute
- $\mathcal{C}$ is a condition over attributes in $\mathcal{A}$
- $\rho \geq 0$ is the maximum allowed deviation
- $\delta \in [0, 1]$ is the violation tolerance ratio

Rule Semantics. For a rule $r = (\mathcal{X} \rightarrow Y, \mathcal{C}, \rho, \delta)$ and dataset $\mathcal{D}$, let $\mathcal{D}_\mathcal{C} = \{t \in \mathcal{D} | t \text{ satisfies } \mathcal{C}\}$ be the subset of tuples satisfying condition $\mathcal{C}$. The rule r holds on $\mathcal{D}$ if there exists a linear model $f : \mathcal{X} \rightarrow Y$ that satisfies Eq. 1:

$$\frac{|\{t \in \mathcal{D}_\mathcal{C} || f(\mathcal{X}(t)) - Y(t)| \leq \epsilon\}|}{|\{t \in \mathcal{D}_\mathcal{C}\}|} \geq 1 - \delta \tag{1}$$

where $\mathcal{X}(t)$ and $Y(t)$ denote the projection of tuple t onto attributes $\mathcal{X}$ and Y respectively.

This formulation allows for δ proportion of tuples satisfied $\mathcal{C}$ to violate the ϵ-deviation constraint, making the rule robust to noise and outliers while maintaining statistical significance.

3.2 Mining ACRR on Clean Data

The ACRR mining algorithm consists of three main phases: predicate space construction, condition table generation, and model training with rule filtering. Algorithm 1 outlines the overall process.

Predicate Space Construction. The predicate space construction phase aims to identify attribute value ranges that exhibit strong correlations with target attributes while maintaining computational efficiency.

Correlation-Based Pruning: For each pair of attributes (A_i, A_j), we compute the Pearson correlation coefficient as shown in Eq. 2:

$$\rho_{ij} = \frac{\sum_{k=1}^{n}(a_i^k - \bar{a_i})(a_j^k - \bar{a_j})}{\sqrt{\sum_{k=1}^{n}(a_i^k - \bar{a_i})^2}\sqrt{\sum_{k=1}^{n}(a_j^k - \bar{a_j})^2}} \tag{2}$$

Only attribute pairs with $|\rho_{ij}| \geq \rho_{threshold}$ are retained for predicate generation, where $\rho_{threshold}$ is a user-specified threshold typically set to 0.2 based on empirical analysis.

Algorithm 1. ACRR Mining on Clean Data

Require: Dataset $\mathcal{D}$, parameters δ, $\rho_{threshold}$, $min_{coverage}$
Ensure: Set of discovered rules $\mathcal{R}$

$\mathcal{P} \leftarrow$ ComputePredicateSpace($\mathcal{D}$, $\rho_{threshold}$)
$\mathcal{CT} \leftarrow$ GenerateConditionTable($\mathcal{P}$)
$\mathcal{R} \leftarrow \emptyset$
for each condition $\mathcal{C} \in \mathcal{CT}$ **do**
  $\mathcal{D}_\mathcal{C} \leftarrow \{t \in \mathcal{D} | t \text{ satisfies } \mathcal{C}\}$
  if $|\mathcal{D}_\mathcal{C}| \geq min_{coverage}$ **then**
    for each target attribute $Y \in \mathcal{A}$ **do**
      for each source set $\mathcal{X} \subseteq \mathcal{A} \backslash \{Y\}$ **do**
        $f \leftarrow$ TrainLinearModel($\mathcal{D}_\mathcal{C}$, $\mathcal{X}$, Y)
        if ValidateRule(f, $\mathcal{D}_\mathcal{C}$, $\mathcal{X}$, Y, $\rho_{threshold}$, δ) **then**
          compute ρ
          $\mathcal{R} \leftarrow \mathcal{R} \cup \{(\mathcal{X} \rightarrow Y, \mathcal{C}, \rho, \delta)\}$
        end if
      end for
    end for
  end if
end for
return $\mathcal{R}$

Interval-Based Predicate Generation: For each retained attribute A_j, we partition its value domain into intervals using a recursive splitting strategy. Starting with the full range $[min(A_j), max(A_j)]$, we recursively split intervals where the correlation with any target attribute changes significantly.

The splitting criterion for an interval $[v_1, v_2]$ satisfies Eq. 3:

$$\max_{A_k} |\rho(A_j|_{[v_1,v_{mid}]}, A_k) - \rho(A_j|_{[v_{mid},v_2]}, A_k)| > \theta_{split} \tag{3}$$

where $v_{mid} = \frac{v_1+v_2}{2}$ and $\theta_{split} = 0.1$ is the splitting threshold. This process continues until either the correlation difference falls below the threshold or the interval contains fewer than $min_{coverage}$ tuples or reach the maximum number of splits.

Condition Table Generation. The condition table generation phase combines predicates into meaningful conditions while avoiding semantically inconsistent combinations.

Correlation-Based Grouping: We group predicates based on their correlation relationships with target attributes:

- Strongly related Group: Predicates with weak correlations ($|\rho| > 0.9$)
- Positive Group: Other predicates involving attributes positively correlated with the target
- Negative Group: Other predicates involving attributes negatively correlated with the target

Condition Generation Strategy: To maintain semantic consistency and avoid contradictory conditions, we combine wthe predicates of the positive group and the negative group to form different conditions. Specifically, for the strongly correlated group, each predicate constitutes a condition alone.

Model Mining and Rule Filtering. For each generated condition $\mathcal{C}$, we extract the corresponding data subset $\mathcal{D}_\mathcal{C}$ and attempt to discover regression rules. In this article, we use the simplest linear regression as the rule.

A trained model f is accepted as a valid rule if it satisfies both deviation and violation constraints:

1. Coverage Constraint: $|\mathcal{D}_\mathcal{C}| \geq min_{coverage}$
2. Deviation Constraint: The proportion of tuples with prediction error $\leq \rho$ is at least $(1-\delta)|\mathcal{D}_\mathcal{C}|$

3.3 Robust Mining on Noisy Data

Real-world time series data often contains noise in the form of outliers and missing values. We extend the basic ACRR framework with two dedicated modules to handle these data quality issues while maintaining rule discovery effectiveness.

Outlier Detection and Elimination. Outliers can significantly distort regression model parameters and lead to poor rule quality. We implement an iterative outlier elimination strategy during the model training phase.

Outlier Identification Criterion: For a trained model f and data subset $\mathcal{D}_\mathcal{C}$, a tuple t is considered an outlier if satisfies Eq. 4 :

$$|f(\mathcal{X}(t)) - Y(t)| > \gamma \cdot \rho \tag{4}$$

where $\gamma = 2$ is the outlier multiplier. This criterion identifies tuples with prediction errors significantly exceeding the acceptable deviation threshold.

Iterative Elimination Process: Algorithm 2 details the iterative outlier removal process:

The iterative process continues until no more outliers are detected or a maximum number of iterations ($max_{iter} = 3$) is reached. This prevents over-elimination while ensuring convergence.

Rule-Guided Missing Value Imputation. We adopt an iterative imputation strategy based on rule feedback. Initially, rules are mined from complete tuples. Then, these rules are applied to estimate missing values in incomplete records if their known values satisfy the rule condition. Predicted values are averaged from all applicable rules. The updated dataset is used to re-mine rules, and the process continues until the number of complete tuples stabilizes.

This integration of rule mining with imputation ensures that the filled values align with the discovered constraints, preserving consistency and improving downstream rule discovery.

Algorithm 2. Iterative Outlier Removal

Require: Data subset $\mathcal{D}_C$, source attributes $\mathcal{X}$, target Y, parameters ϵ, γ, max_{iter}
Ensure: Cleaned data subset $\mathcal{D}'_C$, trained model f

$\mathcal{D}_{current} \leftarrow \mathcal{D}_C$
$iter \leftarrow 0$
repeat
  $f \leftarrow$ TrainLinearModel($\mathcal{D}_{current}$, $\mathcal{X}$, Y)
  $outliers \leftarrow \emptyset$
  for each tuple $t \in \mathcal{D}_{current}$ **do**
    if $|f(\mathcal{X}(t)) - Y(t)| > \gamma \cdot \epsilon$ **then**
      $outliers \leftarrow outliers \cup \{t\}$
    end if
  end for
  $\mathcal{D}_{prev} \leftarrow \mathcal{D}_{current}$
  $\mathcal{D}_{current} \leftarrow \mathcal{D}_{current} \backslash outliers$
  $iter \leftarrow iter + 1$
until $|\mathcal{D}_{current}| = |\mathcal{D}_{prev}|$ OR $iter \geq max_{iter}$
return $\mathcal{D}_{current}$, f

Missing Value Estimation: For a tuple t with missing value in attribute Y, we identify all rules $r_i = (\mathcal{X}_i \to Y, \mathcal{C}_i, \epsilon_i, \delta_i)$ such that:

- t satisfies condition $\mathcal{C}_i$ based on its known attribute values
- All attributes in $\mathcal{X}_i$ are observed in tuple t

The imputed value is computed as a weighted average of predictions from applicable rules like Eq. 5:

$$\hat{Y}(t) = \frac{\sum_i w_i \cdot f_i(\mathcal{X}_i(t))}{\sum_i w_i} \tag{5}$$

where w_i is the weight of rule r_i, defined as:

$$w_i = \frac{coverage(r_i) \cdot (1 - violation_{ratio}(r_i))}{max_{coverage} \cdot confidence(r_i)} \tag{6}$$

This weighting scheme prioritizes rules with high coverage, low violation rates, and strong statistical confidence.

Convergence Criterion: The iterative imputation process terminates when the number of complete tuples stabilizes, specifically when:

$$\frac{|Complete(\mathcal{D}^{(k+1)}) \backslash Complete(\mathcal{D}^{(k)})|}{|Complete(\mathcal{D}^{(k)})|} < \tau \tag{7}$$

where $Complete(\mathcal{D}^{(k)})$ denotes the set of complete tuples after iteration k, and $\tau = 0.01$ is the convergence threshold.

3.4 Complexity Analysis

Time Complexity: The overall time complexity of the ACRR mining algorithm is $O(|\mathcal{P}|^k \cdot |\mathcal{A}|^2 \cdot n \cdot m)$, where $|\mathcal{P}|$ is the number of predicates, k is the maximum number of predicates per condition, $|\mathcal{A}|$ is the number of attributes, n is the number of tuples, and m is the cost of training a linear regression model.

The predicate pruning strategy reduces $|\mathcal{P}|$ significantly, while the condition generation constraints limit the exponential growth in the number of conditions, making the algorithm practical for real-world datasets.

Space Complexity: The space complexity is $O(|\mathcal{P}|^k + |\mathcal{R}| \cdot |\mathcal{A}|)$, dominated by storing the condition table and discovered rules. The iterative modules add minimal space overhead as they operate on subsets of the original data.

This complexity analysis demonstrates that our algorithm scales reasonably with dataset size and dimensionality, particularly when combined with the correlation-based pruning strategy.

4 Experiments

To evaluate the effectiveness and robustness of our proposed ACRR mining algorithm, we conduct comprehensive experiments on three real-world datasets. We compare the rule mining results under different settings, analyze the impact of pruning strategies and noise-tolerant mechanisms, and benchmark against existing methods.

4.1 Datasets

We use the following three numerical or time series datasets:

- **AirQuality** [6]: A time series dataset collected from gas sensors deployed in a polluted urban area in Italy. It records hourly measurements of gas concentrations and sensor responses from March 2004 to February 2005. Size: 9357 rows, 13 columns.
- **MNPP** [11]: A numerical dataset simulating gas turbine behavior in a Combined Diesel-Electric and Gas (CODLAG) propulsion system of a naval vessel. Size: 11,934 rows, 18 columns.
- **YFJ**: A time series dataset from an induced draft fan system, recording temperatures and pressures of various components. We randomly sample 20,000 rows and 12 columns from the original dataset (over 1 million rows) for efficiency.

4.2 Effect of Predicate Pruning

The predicate pruning mechanism plays a crucial role in balancing computational efficiency and rule completeness in ACRR mining. We systematically evaluate the impact of different correlation thresholds ρ on both algorithm performance and rule discovery quality across multiple datasets.

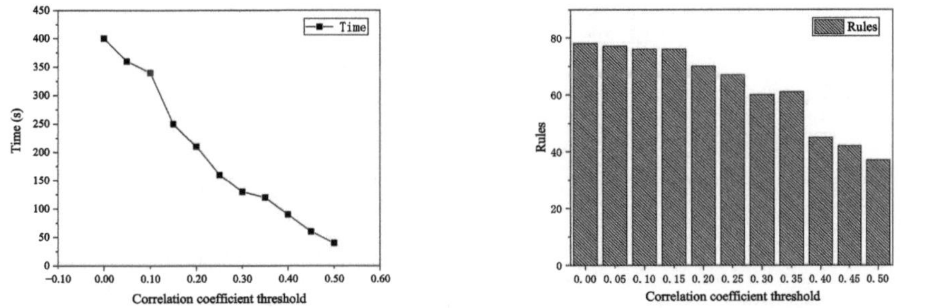

(a) Runtime vs. correlation threshold (b) Rule count vs. correlation threshold

Fig. 2. Performance comparison on AirQuality dataset.

Experimental Setup: We vary the correlation threshold ρ from 0.0 to 0.5 with increments of 0.05, measuring runtime and the number of discovered rules. For each threshold, we maintain consistent parameters: violation tolerance $\delta = 0.001$. Each experiment is repeated 5 times to ensure statistical reliability.

Results on AirQuality Dataset: On the AirQuality dataset (Fig. 2a and 2b), we observe a dramatic reduction in runtime as ρ increases. At $\rho = 0.1$, the algorithm requires approximately 350 s, which drops exponentially to under 250 s at $\rho = 0.4$. This performance improvement is attributed to the significant reduction in predicate space.

Regarding rule discovery, the algorithm maintains stable performance with 70–80 rules discovered when $\rho \leq 0.2$. However, beyond $\rho = 0.4$, rule count drops precipitously to fewer than 40 rules, indicating over-aggressive pruning that eliminates meaningful weak correlations. Statistical analysis shows that the optimal balance occurs at $\rho = 0.2$, where runtime is reduced by 50% while preserving 95% of discoverable rules.

Results on YFJ Dataset: The YFJ dataset exhibits similar trends (Fig. 3a and 3b). Runtime reduction decreases from over 50 s at $\rho = 0.1$ to approximately 30 s at $\rho = 0.3$. The rule count remains relatively stable around 200–250 rules for $\rho \leq 0.2$, then shows gradual decline.

Correlation Analysis: We perform statistical analysis between pruning thresholds and rule quality metrics. The Pearson correlation coefficient between ρ and average rule R^2 score is negative, nearly -0.73, indicating that higher thresholds lead to rules with better model fit but fewer discoveries. Cross-dataset analysis confirms $\rho = 0.2$ as the optimal threshold, achieving 2–3x speedup while maintaining over 90% rule retention.

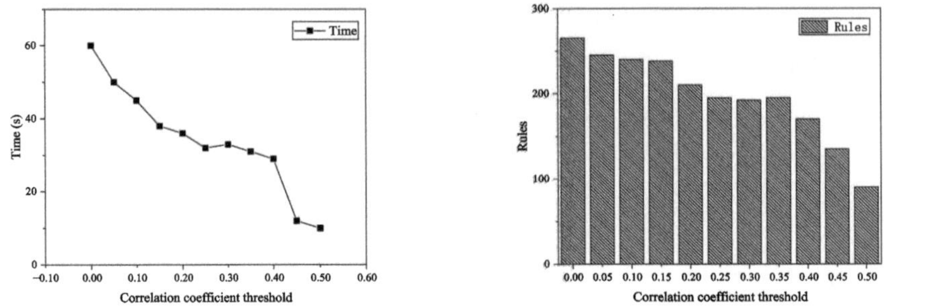

(a) Runtime vs. correlation threshold (b) Rule count vs. correlation threshold

Fig. 3. Performance comparison on YFJ dataset.

4.3 Effect of Violation Tolerance

The violation tolerance parameter δ fundamentally determines the algorithm's ability to discover approximate relationships in noisy data. We conduct comprehensive experiments to understand its impact on rule discovery patterns and quality across different data characteristics.

Experimental Design: We systematically vary δ from 0.0 (exact rules only) to 0.03 (3% violation tolerance). For each setting, we categorize discovered rules into exact rules ($\delta = 0$) and approximate rules ($\delta > 0$), analyzing their distribution, statistical significance, and practical interpretability.

YFJ Dataset Analysis: The YFJ dataset, characterized by industrial sensor measurements with inherent noise, demonstrates the critical importance of violation tolerance (Fig. 4a). At $\delta = 0$, only 6 exact rules are discovered, severely limiting the algorithm's utility for real-world applications. As δ increases, approximate rule discovery grows exponentially, reaching 350 rules at $\delta = 0.02$.

Qualitative analysis reveals that approximate rules in YFJ capture meaningful industrial relationships that exact rules miss due to measurement uncertainties and sensor drift. For instance, temperature-pressure relationships in thermal systems exhibit natural variations (typically 2–5%) that strict exact rules cannot accommodate, but approximate rules with $\delta = 0.01$ successfully identify.

MNPP Dataset Analysis: The MNPP dataset, derived from controlled simulations, exhibits similar behavior (Fig. 4b). The dataset yields 194 exact rules at $\delta = 0$, indicating high data quality and strong deterministic relationships. As δ increases, approximate rule growth is modest, reaching 370 rules at $\delta = 0.03$. This shows that increasing the tolerance for violations can help to uncover potential rules.

Rule Quality Assessment: Results show that There is no optimal solution for the value of tolerance violation. It depends on the degree of control that the researcher wants to achieve. If more precise rules are needed, the violation tolerance can be set smaller, otherwise it can be set larger.

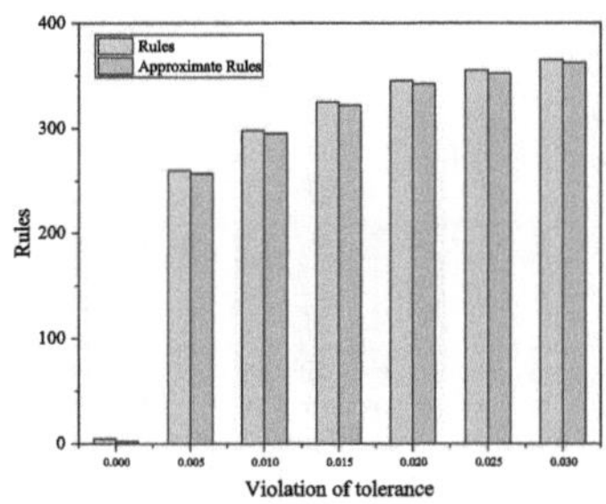

(a) Rule count vs. violation tolerance on YFJ

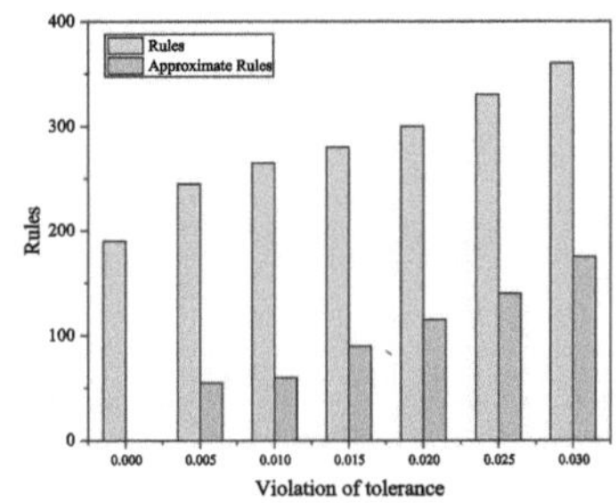

(b) Rule count vs. violation tolerance on MNPP

Fig. 4. Rule count vs. violation tolerance.

4.4 Comparison with CRR

We conduct a comprehensive comparison between our ACRR algorithm and the existing Conditional Regression Rule (CRR) mining method [12] to demonstrate the advantages of approximate rule mining.

Experimental Configuration: We implement both algorithms with identical preprocessing and use the AirQuality dataset with varying tuple sizes to assess scalability. For fair comparison, we set parameters to their optimal values: The correlation coefficient threshold is set to 0.1, and the violation tolerance is set to 0.001 based on the size of the dataset.

Runtime Comparison: Figure 5a shows that CRR consistently outperforms ACRR in terms of runtime, requiring 20–40% less time across all dataset sizes. This advantage stems from CRR's stricter constraints, which result in fewer candidate conditions and faster model validation. However, the runtime difference remains manageable, with ACRR requiring less than 5 min even on the largest dataset.

Rule Discovery Comparison: The crucial difference lies in rule discovery capability (Fig. 5b). ACRR discovers 1–2 times more rules than CRR across all dataset sizes. For the largest AirQuality dataset (9000+ tuples), CRR identifies

55 rules while ACRR discovers 90 rules. Manual inspection reveals that the additional rules discovered by ACRR capture meaningful approximate relationships that CRR misses due to its strict deviation requirements.

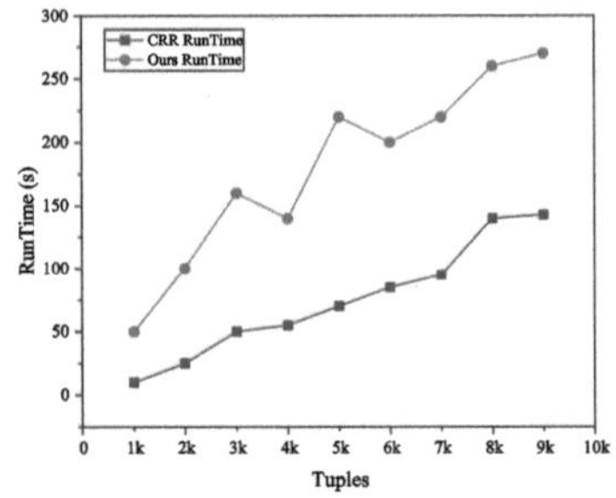

(a) Runtime comparison: ACRR vs. CRR on AirQuality

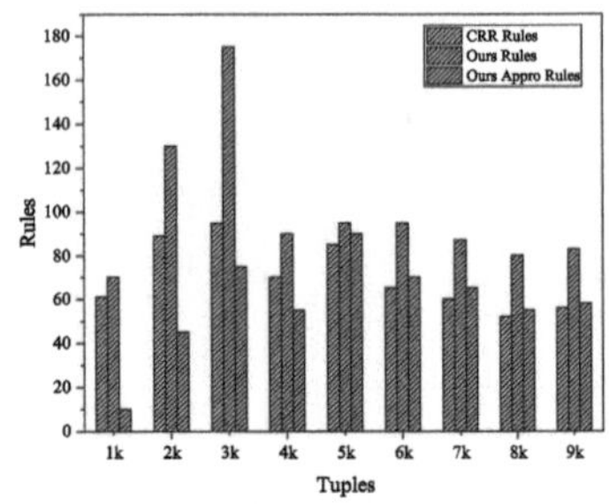

(b) Rule count comparison: ACRR vs. CRR on AirQuality

Fig. 5. Comparison between ACRR and CRR on AirQuality dataset.

4.5 Impact of Outlier Removal

The outlier handling mechanism is essential for robust rule mining in noisy time series data. We evaluate its effectiveness across different datasets and analyze its impact on both exact and approximate rule discovery.

Outlier Detection Strategy: Our iterative outlier removal process identifies tuples with prediction errors exceeding 2ϵ during model training. The algorithm iteratively removes outliers and retrains models until convergence (error change < 0.001) or maximum iterations are reached. We compare results with and without this module to quantify its impact.

Quantitative Results: Table 1 presents comprehensive results across all three datasets. The outlier filtering consistently improves rule discovery, with the most significant impact on approximate rules. On AirQuality, approximate rules increase from 74 to 114 (54% improvement). This pattern reflects the algorithm's enhanced ability to identify underlying relationships when extreme values are filtered out.

Dataset-Specific Analysis: AirQuality shows the largest improvement, with total rules increasing from 98 to 138. Analysis reveals that environmental sensors are particularly susceptible to temporary spikes due to localized events (e.g., vehicle emissions), which the outlier removal effectively handles.

MNPP demonstrates more modest improvements (203 to 205 total rules) due to its simulation origin and inherently cleaner data. However, the few additional rules discovered are highly significant, capturing subtle nonlinear effects.

YFJ interestingly shows a slight decrease in approximate rules (233 to 217), suggesting that some apparent outliers actually represent valid operational states in industrial systems. This highlights the importance of domain knowledge in outlier detection.

Performance Impact Analysis: Outlier removal adds approximately 15–25% computational overhead due to iterative model retraining. However, this cost is justified by the improved rule quality and discovery rate.

Table 1. Effect of outlier filtering on rule count.

Dataset	Exact.Rules	Approx.Rules	Filter
AirQuality	24	74 → 114	Yes
MNPP	194	9 → 11	Yes
YFJ	6	233 → 217	Yes

4.6 Missing Value Imputation

Our rule-guided imputation strategy represents a novel approach that integrates rule discovery with missing value estimation. We comprehensively evaluate its effectiveness under various missing data scenarios and compare it with established imputation methods.

Experimental Design: We artificially introduce missing values at rates from 1% to 30% using Missing Completely At Random (MCAR) mechanism to ensure unbiased evaluation. For each missing rate, we measure both data error(similar to RMSE) and rule discovery performance (rule count and quality).

Performance Under Varying Missing Rates: Figure 6a and 6b demonstrate the algorithm's robustness to missing data. Rule count remains remarkably stable for missing rates up to 10%, with only 8% decrease compared to complete data. Beyond 15% missing rate, performance degrades more significantly, though the algorithm continues to discover meaningful rules even at 30% missing rate (8 rules vs. 78 for complete data).

When the missing rate is below 30%, the data error is not much different, with only a small fluctuation, but there is a sign of increasing as the missing rate increases.

Comparative Analysis: Table 2 presents a comprehensive comparison with four established imputation methods under 10% missing rate:

- Mean Imputation: Simple but distorts data distribution, resulting in fewer discoverable rules (91 vs. 117).
- MissForest: Shows competitive imputation accuracy (0.162 RMSE) but discovers fewer rules (96) due to distribution shifts.
- IIM (Iterative Imputation Method): Demonstrates good balance with 106 rules and 0.158 RMSE.
- GAIN: Achieves lowest imputation error (0.150) but moderate rule discovery (105 rules).
- **Our Method**: Achieves highest rule count (117) while maintaining competitive imputation accuracy (0.153 RMSE).

Rule-Guided Imputation Mechanism: Our approach's superiority stems from its iterative feedback loop between rule discovery and imputation. Initially discovered rules from complete tuples guide the imputation of missing values, ensuring consistency with underlying data relationships. This process iterates until convergence, typically requiring 3–5 iterations. The method averages predictions from multiple applicable rules, providing more robust estimates than single-model approaches.

Computational Complexity: The rule-guided imputation adds moderate computational overhead, increasing total runtime by approximately 20–40% compared to mining on complete data. However, this cost is offset by the improved data quality and rule discovery performance, making it worthwhile for applications requiring high-quality rule mining results.

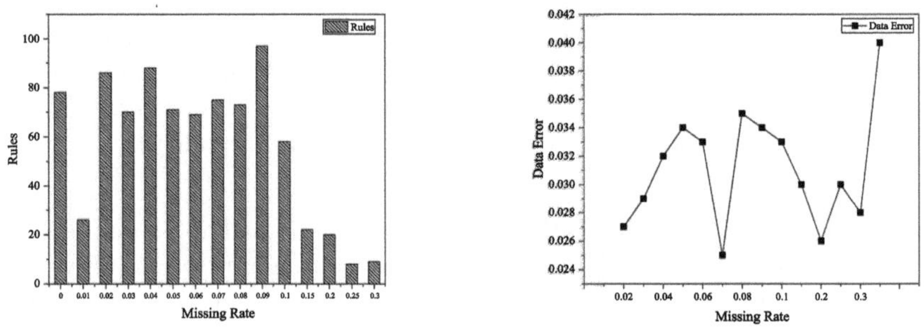

(a) Rule count vs. missing rate

(b) Imputation error vs. missing rate

Fig. 6. Performance under varying missing rates on AirQuality dataset.

Table 2. Rule count comparison under 10% missing rate.

Method	Rules	Imputation Error
Mean	91	0.189
MissForest	96	0.162
IIM	106	0.158
GAIN	105	0.150
Ours	**117**	**0.153**

5 Conclusion

This paper proposes Approximate Conditional Regression Rules (ACRR) and an integrated mining framework to address the challenge of robust regression rule mining in noisy time series data. Our key contributions include: (1) the introduction of ACRR, which extends traditional regression rules with violation tolerance to effectively capture approximate relationships in noisy environments; (2) an efficient mining algorithm incorporating predicate pruning, condition generation, and model validation; and (3) integrated modules for outlier detection and rule-guided missing value imputation, ensuring consistency between data cleaning and rule discovery.

Experimental results on three real-world datasets show that our approach discovers 2–3 times more rules than existing methods while maintaining comparable predictive accuracy. The violation tolerance mechanism proves particularly effective for industrial data, where exact methods often fail to identify meaningful patterns. Furthermore, the proposed outlier handling strategy enhances both the robustness and comprehensiveness of rule discovery.

References

1. Altman, N.: An introduction to kernel and nearest-neighbor nonparametric regression. Am. Stat. **46**(3), 175–185 (1992)
2. Bleifuß, T., et al.: Hydra: a hybrid rule discovery approach for denial constraints. In: Proceedings of the 2017 ACM International Conference on Management of Data (SIGMOD), pp. 1199–1214 (2017)
3. Buuren, S., Groothuis-Oudshoorn, K.: Mice: multivariate imputation by chained equations in R. J. Stat. Softw. **45**(3), 1–67 (2010)
4. Campos, E.C., et al.: BFASTDC: scalable discovery of denial constraints via indexing and pruning. In: Proceedings of the ACM SIGMOD Conference (2021)
5. Chu, X., Ilyas, I.F., Koutris, P.: Discovering denial constraints. In: Proceedings of the VLDB Endowment, vol. 6, pp. 1498–1509 (2013)
6. De Vito, S., Massera, E.: Air quality dataset (2008). https://archive.ics.uci.edu/ml/datasets/Air+Quality
7. Fan, W., Geerts, F., Jia, X., Kementsietsidis, A.: Conditional functional dependencies for capturing data inconsistencies. ACM Trans. Database Syst. (TODS) **33**(2), 1–48 (2008)

8. Huhtala, Y., Kärkkäinen, J., Porkka, P., Toivonen, H.: Tane: an efficient algorithm for discovering functional and approximate dependencies. Comput. J. **42**, 100–111 (1999)
9. Papenbrock, T., Naumann, F.: A hybrid approach for efficient unique column combination discovery. In: Proceedings of the 31st IEEE International Conference on Data Engineering (ICDE), pp. 312–323 (2015)
10. Phung, T.T., Tran, T.: Midae: a multi-input denoising autoencoder for missing data imputation. arXiv preprint arXiv:1905.06843 (2019)
11. Repository, U.: Naval propulsion plant condition monitoring dataset (MNPP) (2014). https://archive.ics.uci.edu/ml/datasets/Condition+monitoring+of+hydraulic+systems
12. Song, S., Zhang, T., Fan, W.: Conditional regression rules: towards discovering reliable patterns from noisy data. Proc. VLDB Endowment **15**(12), 3421–3434 (2022)
13. Stekhoven, D.J., Bühlmann, P.: Missforest—non-parametric missing value imputation for mixed-type data. Bioinformatics **28**(1), 112–118 (2012)
14. Wyss, C., Giannella, C., Robertson, E.: FastFDs: a heuristic-driven, depth-first algorithm for mining functional dependencies from relation instances extended abstract. In: Kambayashi, Y., Winiwarter, W., Arikawa, M. (eds.) DaWaK 2001. LNCS, vol. 2114, pp. 101–110. Springer, Heidelberg (2001). https://doi.org/10.1007/3-540-44801-2_11
15. Yoon, J., Jordon, J., van der Schaar, M.: Gain: missing data imputation using generative adversarial nets. In: Proceedings of the 35th International Conference on Machine Learning (ICML), pp. 5689–5698 (2018)

Mobile Sensing and Computing

AUV Cluster Obstacle Avoidance Algorithm Based on DHMATD3

Yu Meng(✉) and Yuxi Wu

Harbin Engineering University, Harbin 150001, China
hrbeum@126.com

Abstract. In response to the suboptimal obstacle avoidance performance of Autonomous Underwater Vehicle clusters (AUV clusters) and the issues of slow convergence speed and poor network performance associated with MATD3 networks, we propose a novel AUV cluster obstacle avoidance algorithm based on DHMATD3. By computing the sliding window average of the iterative TD error during the MATD3 network training process, we dynamically adjust the delayed update frequency of the MATD3 network. Subsequently, we utilize a GRU network to process historical sample data and incorporate it into the experience pool, thereby altering the sampling method of the MATD3 network experience pool to expedite the learning of excellent data. Given that certain outputs of the GRU network may contain irrelevant information, we introduce an attention mechanism to ensure that the GRU network focuses solely on useful information, thereby optimizing the structure of the MATD3 network. We apply this approach to address AUV cluster obstacle avoidance problems. Experimental results demonstrate that this algorithm effectively enhances the performance and convergence speed of the MATD3 network, leading to more optimal AUV cluster obstacle avoidance decisions and increased stability.

Keywords: Autonomous Underwater Vehicle clusters · Obstacle Avoidance · MATD3 network

1 Introduction

AUVs execute various tasks in complex marine environments, facing obstacles of different states during navigation. Successfully avoiding these obstacles and completing tasks rely heavily on obstacle avoidance technology. Consequently, obstacle avoidance techniques for AUVs are among the focal points of research in the field of AUVs [1]. The obstacle avoidance methods for AUVs can be categorized into two types based on the type of obstacles: global obstacle avoidance for known static obstacles and local obstacle avoidance for unknown dynamic obstacles [2]. In environments with known static obstacles, the common approach involves modeling the entire environment. AUVs then use algorithms to plan a feasible path based on this global model and navigate along it. Common methods for handling known static obstacles include the A* algorithm [3], particle swarm optimization [4], and ant colony optimization [5]. Peng [6] addressed the weak disturbance resistance of the PSO-GA algorithm by introducing an anti-interference

X. Zhou et al. (Eds.): GPC 2024, LNCS 15225, pp. 151–166, 2026.
https://doi.org/10.1007/978-981-95-1346-8_10

and fault-tolerant processing algorithm designed for MEMS inertial navigation systems. This enhancement led to a more stable AUV obstacle avoidance planning. Jia et al. [7] proposed an improved particle swarm optimization algorithm, incorporating an optimal particle "mutation" process to enhance the algorithm's solving capability. They integrated the velocity obstacle method to achieve a novel path planning algorithm for both global static and local dynamic environments. K. T. et al. [8] optimized the AUV's trajectory during motion by refining the cost function of the traditional genetic algorithm (GA), thereby reducing energy consumption. In unknown dynamic underwater environments, it is often difficult, if not impossible, to directly obtain global information containing various obstacles. In such scenarios, AUV obstacle avoidance requires a real-time controller to navigate around these unknown dynamic obstacles. Common methods for implementing such a controller include Artificial Potential Field (APF), Rapidly-exploring Random Tree (RRT) [9], which are sensor-based approaches relying on AUV's own sensors, and Reinforcement Learning (RL) [10], Deep Reinforcement Learning (DRL) [11], which are machine learning-based artificial intelligence decision methods. Xi et al. [12] proposed a method based on the Double-layer Deep Deterministic Policy Gradient (DDPG) algorithm, which enhances stability and sampling efficiency during training by employing two layers of DDPG networks and the Hindsight Experience Replay (HER) method. Xing et al. [13] integrated the Improved Artificial Potential Field (IAPF) method with Deep Reinforcement Learning (DRL) for path planning. They optimized the reward function in the DRL algorithm and utilized the generated paths to optimize future trajectories, thereby achieving AUV obstacle avoidance.

The Twin Delayed Deep Deterministic Policy Gradient (TD3) algorithm is a relatively novel Deep Reinforcement Learning (DRL) technique. It was introduced by Fujimoto in 2018 as an enhancement of the Deep Deterministic Policy Gradient (DDPG) algorithm, aiming to address some of the shortcomings present in DDPG [14]. Similar to the DDPG network, the TD3 network also has a variant suitable for multi-agent systems. This variant is called the Multi-Agent Twin Delayed Deep Deterministic Policy Gradient (MATD3) algorithm, which is based on the TD3 network architecture. Sarkar et al. [15] proposed a weighted loss function to enhance the update mechanism of the Critic network's loss function in the MATD3 network. This weighted loss function is applied to each Agent's Critic network, aiming to improve the performance and stability of the network. Fan et al. [16] introduced a multi-robot Twin Attention Reinforcement Learning method, which incorporates twin attention critics and delayed decay strategies. This approach enhances the robustness of the MATD3 network. The TD3 network is commonly utilized for addressing continuous control problems such as robot control and autonomous driving, showcasing its efficacy in learning policies within complex environments. Compared to other algorithms, TD3 networks offer advantages such as strong stability, robust generalization capabilities, wide applicability, and shorter model training times. These attributes provide TD3 networks with a natural advantage in addressing AUV obstacle avoidance problems within complex environments. Indeed, despite its strengths, the TD3 method does have drawbacks such as high computational requirements, elevated learning costs, and long training times.

In summary, this paper utilizes MATD3 as the foundational network to develop a novel AUV cluster obstacle avoidance algorithm based on DHMATD3. This algorithm is then applied to AUV cluster obstacle avoidance scenarios. The main novelties and contributions of this article are as follows:

1. Addressing the issue of suboptimal solutions in the later stages of network training due to a fixed dynamic delay update strategy, a dynamic delay update optimizer is introduced. This optimizer dynamically adjusts the update frequency of the TD3 network during training by continuously computing the TD error, ensuring that the network can access better data earlier in the training process.
2. To tackle the problem of underutilization of excellent sample data in the replay experience pool of the DDPG network caused by random sampling, an Attention-based GRU network layer is utilized to analyze historical sample data. This layer selects outstanding sample data and assigns them higher priority when adding them to the experience pool, thereby increasing the likelihood of utilizing excellent data and accelerating the network training process.

2 Related Works

2.1 AUV Digital Model

In consideration of the variation in the detection range of the AUV in real environments, which depends on the depth difference between the target and the AUV itself, this paper simplifies the environment by only considering AUV obstacle avoidance research in dynamic environments with constant water depth. Specifically, the study focuses on the two-dimensional plane of AUV active obstacle avoidance. For ease of representation, the AUV is abstracted as a circular shape with dimensions of 1 * 1 unit. The obstacle avoidance system of the AUV primarily relies on the forward sonar it carries. Assuming the sonar detection area forms a fan-shaped region with a lateral cone angle of 100° and a maximum detection distance of 10m, the vertical detection range of the sonar is not considered. The Cartesian coordinate system for the AUV kinematic model is depicted in Fig. 1.

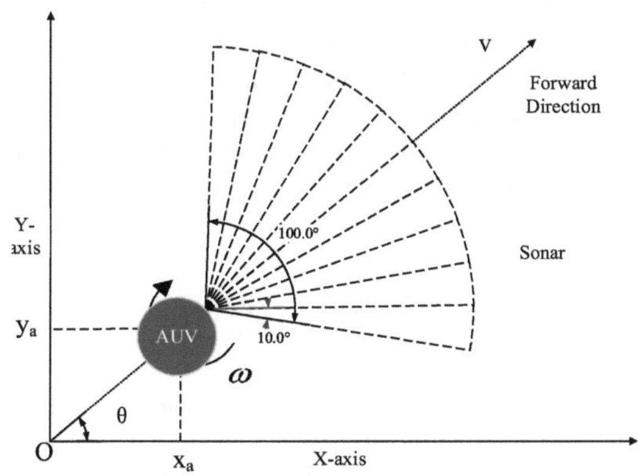

Fig. 1. AUV kinematic model

Establishing a Cartesian coordinate system with O as the origin, the AUV is represented by a red circle, with its centroid serving as the coordinate representation. Thus, the AUV's position can be denoted as (x_a, y_a), where x represents the position along the horizontal axis (x-axis), and y represents the position along the vertical axis (y-axis). The motion of the AUV can be described by the following parameters: θ denotes the heading angle of the AUV relative to the horizontal axis (x-axis). ω represents the angular velocity of the AUV, i.e., the yaw rate. v denotes the forward velocity of the AUV.

During the navigation process towards a target point, an AUV exhibits three primary behaviors: cruising, obstacle avoidance and heading towards the target. The main actions performed by the AUV during navigation include moving forward, turning, and stopping/waiting. Forward movement can be either at a constant speed or with constant acceleration, while turning, under constant water depth conditions, can be categorized into left or right turns. In the obstacle avoidance process, the AUV achieves these behaviors primarily by dynamically adjusting the magnitude of the yaw rate and forward velocity. Therefore, this paper considers forward velocity v and yaw rate ω as the fundamental actions $a = [v, \omega]$.

If the coordinates of the AUV's location at time (x_t, y_t) are known, with a time interval of Δt, and the AUV's heading angle is θ, then the heading angle of the AUV θ' at time $t + \Delta t$ is given by Eq. (1). By substituting this into Eq. (2), we can obtain the position of the AUV $(x_{t+\Delta t}, y_{t+\Delta t})$ at time $t + \Delta t$.

$$\theta' = \theta + \omega \Delta t \tag{1}$$

$$\begin{cases} x_{t+\Delta t} = x_t + v * \Delta t * \cos(\theta') \\ y_{t+\Delta t} = y_t + v * \Delta t * \sin(\theta') \end{cases} \tag{2}$$

2.2 MATD3

MATD3 adopts the architecture of Centralized Twin Delayed Deep Deterministic Policy Gradient (CTDE), with a "centralized" Critic network consistent with TD3, employing a structure of dual Critic networks. The network architecture diagram is as shown in Fig. 2. The minimum value between the two Q-values is chosen. The target Q-value of the MATD3 network is represented by Eq. (3), while the two Critic networks have two loss functions as shown in Eq. (4). Noise is introduced for each agent to smooth out policy selection, as indicated in Eq. (5). The final target Q-value calculation formula is illustrated in Eq. (6).

$$Q_{\text{target}} = r + \gamma \min_{i=1,2;j=1,\ldots\ldots,n} Q^{j}_{\varphi'_i}\left(s_{t+1}, \pi_{\theta'_j}(s_{t+1})\right) \tag{3}$$

$$\begin{cases} L\left(\varphi^{Q}_{1,i}\right) = E_{s_t \sim \rho^{\beta}, a_t \sim \beta, r_t \sim E}\left[\left(Q_i\left(s_t, a_t^1, \ldots\ldots, a_t^n \middle| \theta^{Q}_{1,i}\right) - Q_{t\arg et}\right)^2\right] \\ L\left(\varphi^{Q}_{2,i}\right) = E_{s_t \sim \rho^{\beta}, a_t \sim \beta, r_t \sim E}\left[\left(Q_i\left(s_t, a_t^1, \ldots\ldots, a_t^n \middle| \theta^{Q}_{2,i}\right) - Q_{t\arg et}\right)^2\right] \end{cases} \tag{4}$$

$$a_i = \pi_{\theta_i}(s) + clip(\delta, -c, c) \tag{5}$$

$$Q_{\text{target}} = r + \gamma \min_{i=1,2;j=1,\cdots\cdots,n} Q^{j}_{\varphi_i'}\Big(s_{t+1}, clip\Big(\pi_{\theta_j'}(s_{t+1}) + clip(\delta, -c, c), a_{low}, a_{high}\Big)\Big) \tag{6}$$

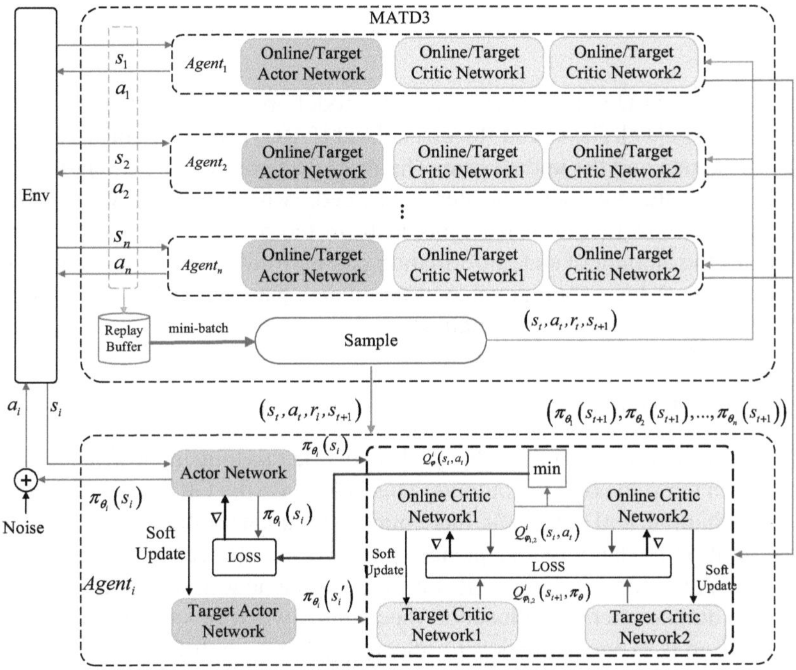

Fig. 2. MATD3 Network Architecture Diagram

2.3 Moving Average

The moving average is obtained by sliding a fixed-size window over a data sequence and calculating the average of the data within the window, thereby obtaining a series of non-linear average values. The specific implementation of the moving average is shown in Fig. 3.

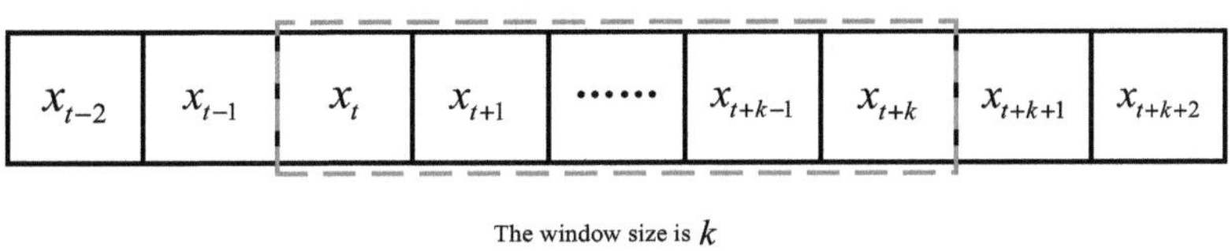

Fig. 3. Illustration of Moving Average

A moving window of size k is illustrated, where x_t represents the data at time t. The average value of this window is represented by Eq. (7).

$$v_{avg} = \frac{x_t + x_{t+1} + \cdots\cdots + x_{t+k-1} + x_{t+k}}{k} \quad (7)$$

2.4 GRU

The Gated Recurrent Unit (GRU) is a network model optimized from Recurrent Neural Networks (RNNs) [17].The Gated Recurrent Unit (GRU) primarily consists of two gate units: the reset gate and the update gate. The reset gate controls the extent to which the information from the previous time step is ignored. A higher output value of the reset gate implies less information being ignored from the previous time step. The update gate regulates the amount of state information passed from the previous time step to the current time step. A lower output value of the update gate means less state information from the previous time step being retained.

2.5 Attention Mechanism

When feeding all historical data into the GRU network, this long input sequence may lead to a decrease in the value of GRU outputs. This is because the GRU will correspondingly output for each input, and as a result, there may be redundant information in the outputs. This deterioration of the output sequence can adversely affect the performance of the network. The Attention Mechanism is a critical component in deep learning, designed to mimic the human attention allocation mechanism when processing information. Its primary function is to enable models to selectively focus on different parts of the input data, as stated in [18]. By specifying the importance of GRU outputs, it becomes possible to emphasize certain data while disregarding others. This is illustrated in the specific structure shown in Fig. 4.

The transformation relationships are depicted by Eqs. (8) (9) (10) (11).

$$H_t = GRU(X_t) \quad (8)$$

$$u_t = \tanh(\omega H_t + b_\omega) \quad (9)$$

$$\alpha_t = \frac{\exp\left(u_t^{\mathrm{T}} \mu\right)}{\sum_t \exp\left(u_t^{\mathrm{T}} \mu\right)} \quad (10)$$

$$O_t = \alpha_t * H_t \quad (11)$$

In the equations, X_t, H_t, u_t, O_t represent the data input at time t, the GRU output, the attention probability distribution value, and the attention output, ω and μ are learnable matrices, with α_t referred to as the attention weights.

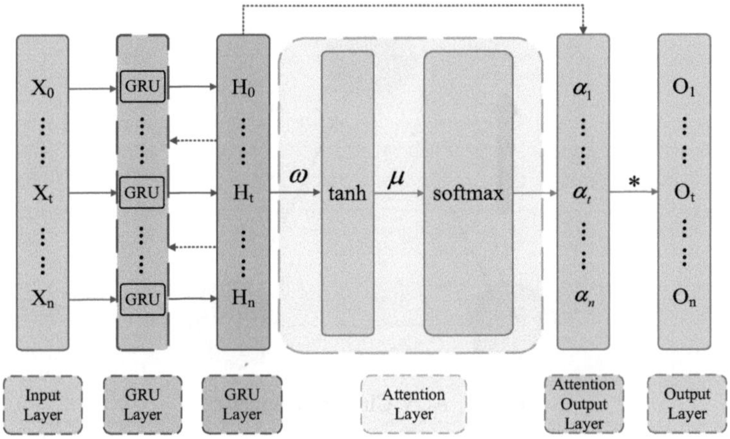

Fig. 4. Attention-based GRU Network

3 Our Approach

3.1 Modeling of AUV Cluster Obstacle Avoidance Behavior

The primary challenge in addressing obstacle avoidance for AUV clusters lies in maintaining the distance between AUVs and obstacles, as well as ensuring the separation between AUVs. The AUV cluster model used in this paper is suitable for scenarios involving obstacle avoidance in constant-depth environments. To facilitate research, the following assumptions are made based on the actual state of AUVs in the environment:

1. AUVs can reliably communicate bidirectionally within their communication range.
2. Other AUVs surrounding each AUV can be detected.
3. AUVs have access to the state information of all other AUVs within their communication range.

To ensure initial stability of the formation, the number of AUVs is set to three, arranged in an equilateral triangle formation. As illustrated in Fig. 5, the three AUVs are placed at the vertices of the triangle, with the centroid located at each vertex. The maximum communication distance for each AUV is the length of the triangle's side.

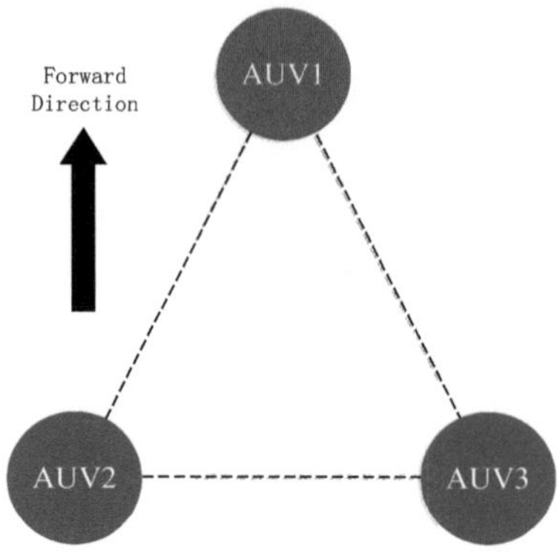

Fig. 5. AUV Cluster Schematic

The state information of AUVs within the cluster now includes not only the AUV's local information but also partial information about other AUVs within the cluster. The motion state S_a of an AUV itself consists of ten equally spaced sonar detection range data, the yaw angle of the AUV towards the target, the forward velocity of the AUV, and the yaw rate of the AUV, as shown in Eq. (12). The distance between the AUV and the target point, the distance between the AUV and obstacles, and the distance between AUVs are denoted as d_t, d_o, d_a, respectively, as shown in Eq. (13).

$$S_a = [S_s, v, \theta, \omega] \tag{12}$$

$$d_{ij}^t = \sqrt{\left(x_i^t - x_j^t\right)^2 + \left(y_i^t - y_j^t\right)^2} \tag{13}$$

In the equations, S_s is an array of length 10, representing sonar detection information. The state information of each AUV in the cluster is represented as Eq. (14).

$$s = [S_a, d_t, d_o, d_a] \tag{14}$$

In the Multi-Agent Deep Reinforcement Learning (MADRL) method, each AUV in the cluster makes decisions independently. The local AUV obstacle avoidance strategy does not need to consider the cluster formation. However, from a holistic perspective, modifications to the cluster formation are required due to changes in terrain. The specific scenario is illustrated in Fig. 6.

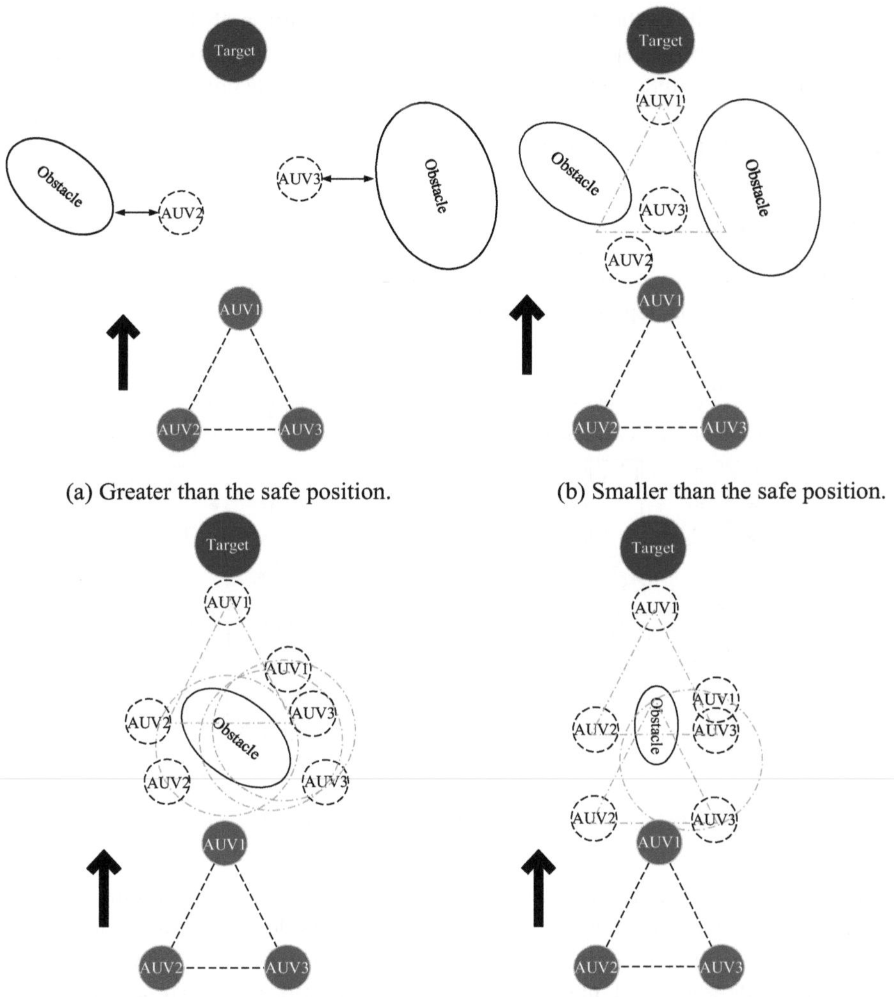

(a) Greater than the safe position. (b) Smaller than the safe position.

(c) Completely obstructing the cluster (d) Partially obstructing the cluster

Fig. 6. Schematic of Cluster Formation Changes for Obstacle Avoidance

3.2 Dynamic Delayed Update Optimizer

In reinforcement learning, TD error can measure the discrepancy between the model's predicted value and the target value at a time step. Therefore, by calculating the TD error of the current network, the training effectiveness of the network can be indirectly understood. The target value v_t is represented as shown in Eq. (15), and the calculation form of TD error is shown in Eq. (16).

$$v_t = r_t + \gamma \cdot \max Q(s_{t+1}, a_{t+1}) \tag{15}$$

$$TD_{error} = v_t - Q(s_t, a_t) \tag{16}$$

The optimizer adjusts the delay update frequency proportionally based on the moving average of the TD error. Specifically, a suitable threshold value is chosen. When the average value is less than the set threshold, the frequency decreases by 1. When the average value exceeds the set threshold, the frequency increases by 1. This process is repeated to modify the delay update frequency.

3.3 DHMATD3

The DHMATD3 network mainly utilizes the MATD3 network, with only partial implementation details depicted in Fig. 7.

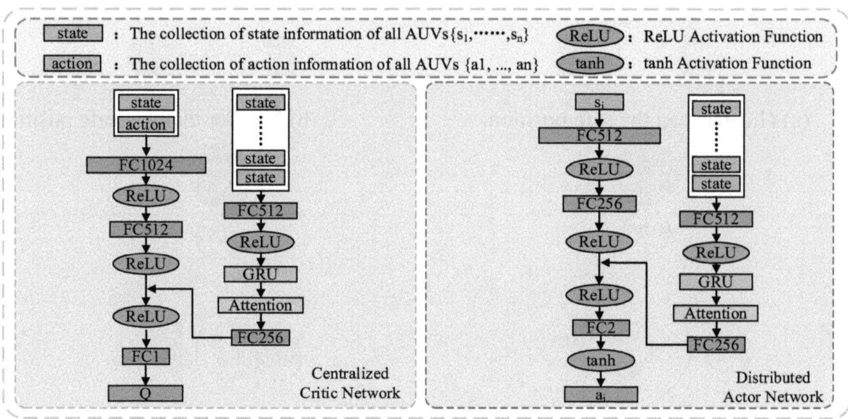

Fig. 7. Schematic Diagram of DHMATD3 Network Architecture

Modify the original structure of the MATD3 network by incorporating the last K historical data into the experience replay pool for sample selection. This implementation realizes the local self-network structure for each AUV within the MATD3 network. Additionally, integrate the dynamic delayed update optimizer to dynamically adjust the update frequency of the Critic network during the training phase. This adjustment aims to obtain better Q-values for updating the Actor network.

The pseudo-code for the DHMATD3 network is presented in Table 1.

Table 1. The pseudo-code for the DHMATD3 network

DHMATD3 network
for AUV_number n = 1 to N do
for episode i = 1 to M do
for step t = 0 to T - 1 do
$a_t^n \leftarrow \pi_{\theta_n}\left(s_t^n\right)+\delta$
$\left(s_t, a_t\right) \leftarrow \left(s_t^1, s_t^2, \ldots, s_t^n ; a_t^1, a_t^2, \ldots, a_t^n\right)$
$\left(r_t, s_{t+1}\right) \leftarrow ENV\left(s_t, a_t\right)$
$\left(o_1, o_2, \ldots, o_n\right) \leftarrow$ GRU-Attention $\left(s_1, s_2, \ldots, s_n\right)$
$SAVE\left(o_1, o_2, \cdots\cdots, o_n\right)$
if reached sufficient training experiences then
$Q_{target} \leftarrow \min\limits_{i=1,2; j=1,\ldots,n} Q_{\varphi_i}^j\left(s_{t+1}\right)+r_t$
$v_{avg} \leftarrow WINDOW\left(\frac{TD_{error_1}+TD_{error_2}}{2}\right)$
$freq' \leftarrow freq$
$\varphi_{i,j} \leftarrow LOSS\left(\varphi_{i,j}^Q\right)\Big\|_{i=1,2; j=1,\ldots,n}$
if $t \ \% \ freq' == 0$ then
$\theta_n \leftarrow \left(s_{t+1}, \pi_{\theta_n}\left(s_{t+1}\right)\right)$
end if
end if
end for
$\left(\theta^{-}、\varphi_1^{-}、\varphi_2^{-}\right) \leftarrow \left(\theta、\varphi_1、\varphi_2\right)$
end for
end for

4 Experiments and Discussion

4.1 Simulation Setup

The experimental setup in this study utilized an Intel Core i5-13490F processor, 32.0 GB RAM, and an RTX 4060Ti graphics card. The simulation environment was constructed using ROS (Robot Operating System) and Gazebo.

4.2 Training Setup

The dataset used in this experiment consists of simulated maps with parameters adjusted. The training map size is 50 m * 50 m, while the testing map size is 100 m * 100 m.

The experiment utilizes the proposed cluster conceptual model with a cluster of 3 AUVs arranged in an equilateral triangle, as depicted in Fig. 8. Figure 8(a) represents the simulated training map for the cluster, while Fig. 8(b) shows the simulated testing map. In both figures, the three red spheres arranged in a triangular formation represent the AUV cluster, and the green area measuring 10 m * 10 m denotes the target area, with green spheres indicating moving obstacles.

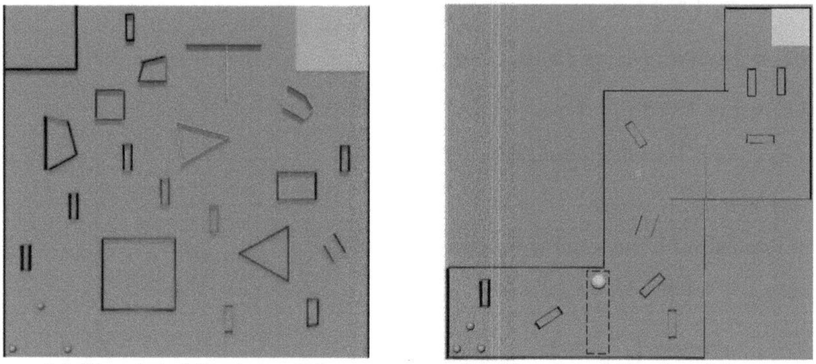

(a) Simulated Training Map (b) Simulated Testing Map

Fig. 8. Cluster Simulation Map

The majority of the network parameters used in MATD3 in this chapter are shown in Table 2.

Table 2. Network Parameters

Parameter	Value
learning rate α	0.01
discount factor γ	0.9999
replay_buffer_*size*	1000000
mini – batch_*size*	32
Max-step T	100000
soft update frequency f	0.01
Max-episode *episode*	100
Max-delay update frequency *freq*	2

The reward function defined for the proposed network in this paper is shown as Eq. (17).

$$r = \begin{cases} r_f, \ d < d_{\min} \\ r_c, \ \text{collision} \\ v - |\omega|, \ \text{other} \end{cases} \tag{17}$$

In the reward function, we choose the parameters as $r_f = 100, r_c = -100, d_{\min} = 1$.

The velocity range for the AUVs used in this paper is 0~1(m/s), and the angular velocity ranges from −1~1(rad/s).

The AUVs are controlled by the ROS system for movement. The stopping criteria for this training include reaching the target point, reaching the maximum number of allowed steps, and colliding with obstacles. To ensure the model's generality, an equal number of obstacles are randomly generated before each training session, with their positions changing. It is ensured that no obstacles appear at the target point and that at least one path exists to reach the target point. After completing one iteration of training, the current network is run 10 times, and the average reward value of these 10 runs is calculated as the final reward for that iteration. The reward values for each iteration are recorded, and the moving average of the reward values for the entire training process is calculated using a sliding window of size 10.

4.3 Experimental Results

To validate the advantages of the MATD3 network in cluster problems and the performance impact of the DHMATD3 network on the MATD3 network, this experiment conducts 400 iterations of training on the preset environment for both the MATD3 and DHMATD3 networks. The reward values are recorded, and reward value and moving average value change graphs are generated, as shown in Fig. 9.

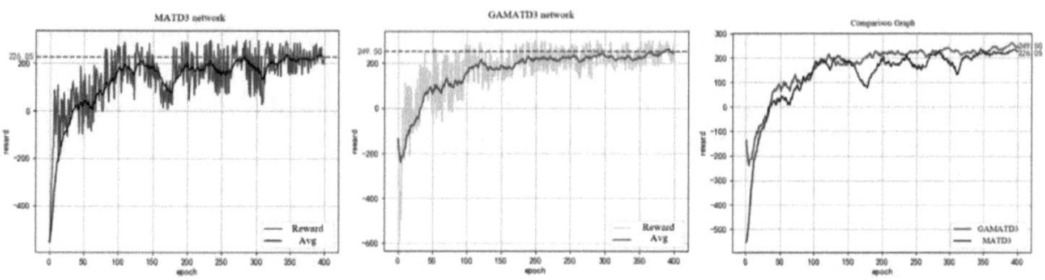

(a)MATD3 network (b)DHMATD3 network (c) Comparison Graph

Fig. 9. MATD3 Network and DHMATD3 Network Reward Iteration Graphs

Analyzing Fig. 9, we can observe that Fig. 9(a) depicts the reward value changes over 400 iterations of training for the MATD3 network, while Fig. 9(b) shows the reward value changes over 400 iterations of training for the DHMATD3 network. Figure 9(c) presents the comparison of the moving average values of the rewards between the two networks.

From the curve variations, it can be seen that the reward value fluctuations of the DHMATD3 network are significantly smaller than those of the MATD3 network. Analyzing the specific values of the rewards, the MATD3 network begins to converge around 250 iterations, and it tends to stabilize around 320 iterations with a stable reward value of 226.05. On the other hand, the DHMATD3 network achieves basic convergence around 200 iterations, with a stable reward value of 249.50. The convergence speed is improved by 25% compared to MATD3, and the moving average value at convergence is increased by 10.37%. Additionally, the average value fluctuations after convergence are significantly smaller than those of the MATD3 network.

To verify the impact of DHMATD3 network on the performance of MATD3 network, conduct 10 rounds of testing on the trained MATD3 network and DHMATD3 network with 400 iterations on the test map. Record the total number of movement steps, collision occurrences, and rewards of the AUV cluster, and calculate the average values of movement steps and rewards, as shown in Table 3.

Table 3. Test results

Network Name	Average Steps	Average Reward	Number of Collisions
MATD3 network	4120	266	0
DHMATD3 network	3750	300	0

Plot the movement trajectories and obstacle avoidance details of both networks on the test map, as shown in Fig. 10.

Analysis of Table 3 reveals that the DHMATD3 network exhibits a slight decrease in average steps compared to the MATD3 network. In terms of the average rewards over 10 tests, the DHMATD3 network performs better, reaching the highest cluster reward value. This indicates that the cluster moves along the optimal route and reaches the target area. However, the MATD3 network encounters situations where it is too close to obstacles or other AUVs, resulting in suboptimal movement and failure to achieve the highest reward value. Neither network experiences collisions in the 10 tests. Examining Fig. 10, we observe that the trajectory of the MATD3 network in Fig. 10(a) shows oscillations due to proximity to obstacles, but it does not follow the optimal route. In contrast, the trajectory of the DHMATD3 network in Fig. 10(b) closely follows the optimal route, resulting in smoother movement and the fewest steps taken.

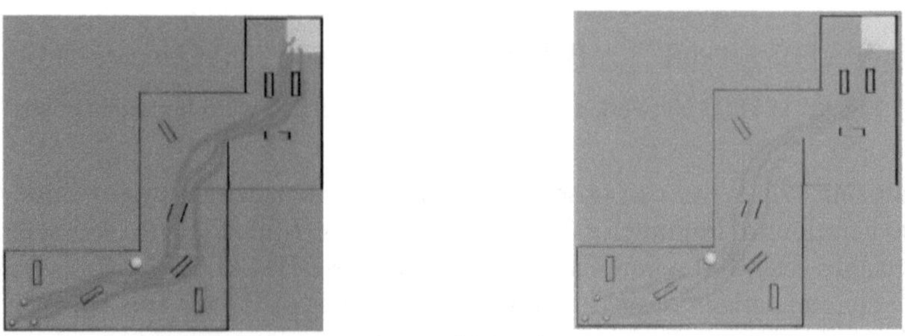

(a) Trajectory plot of MATD3 network (b) Trajectory plot of DHMATD3 network

Fig. 10. Trajectory plots of the test maps

5 Conclusion

The proposed AUV cluster obstacle avoidance algorithm based on DHMATD3 has shown improvements in both network convergence speed and performance compared to the MATD3 network. It has achieved good obstacle avoidance results for the cluster. However, the method presented in this paper only addresses obstacle avoidance in a two-dimensional context, while the actual marine environment is three-dimensional. Therefore, further validation is needed to determine whether this approach is suitable for obstacle avoidance in a three-dimensional context.

References

1. Zhu, G., Shen, Z., Liu, L., et al.: AUV dynamic obstacle avoidance method based on improved PPO algorithm. IEEE Access **10**, 121340–121351 (2022)
2. Cheng, C., Sha, Q., He, B., et al.: Path planning and obstacle avoidance for AUV: a review. Ocean Eng. **235**, 109355 (2021)
3. Sui, F., et al.: ACO+ PSO+ A*: a bi-layer hybrid algorithm for multi-task path planning of an AUV. Comput. Industr. Eng. **175**, 108905 (2023)
4. Shen, Y., Zhang, J., Han, G., et al.: Multi-AUV data retrieval method based on multiple signal classification algorithm. In: 2023 International Conference on Intelligent Communication and Networking (ICN), pp. 73–77. IEEE (2023)
5. Chawra, V.K., Gupta, G.P.: Optimized coverage-aware trajectory planning for AUVs for efficient data collection in underwater acoustic sensor networks. Evol. Intell. **16**(2), 401–416 (2023)
6. Cheng, P.: Working path optimization of AUV manipulator based on PSO-GA algorithm. In: 4th International Conference for Emerging Technology (INCET), pp. 1–6 (2023)
7. Jia, H., Du, F.: The path planning research with global static and local dynamic of the AUV. For. Chem. Rev. 726–740 (2021)
8. Tanakitkorn, K., Wilson, P.A., Turnock, S.R., Phillips, A.B.: Grid-based GA path planning with improved cost function for an over-actuated hover-capable AUV. In: 2014 IEEE/OES Autonomous Underwater Vehicles (AUV), pp. 1–8 (2014)
9. Zhang, X., Zhu, T., Du, L., Hu, Y., et al.: Local path planning of autonomous vehicle based on an improved heuristic bi-RRT algorithm in dynamic obstacle avoidance environment. Sensors **22**(20), 7968 (2022)
10. Chronis, C., Anagnostopoulos, G., Politi, E., Dimitrakopoulos, G., Varlamis, I.: Dynamic navigation in unconstrained environments using reinforcement learning algorithms. IEEE Access **11**, 17984–118001 (2023)
11. Yuan, J., Han, M., Wang, H., Zhong, B., Gao, W., Yu, D.: AUV collision avoidance planning method based on deep deterministic policy gradient. J. Marine Sci. Eng. **11**(12), 2258 (2023)
12. Lyu, X., Sun, Y., Wang, L., Tan, J., Zhang, L.: End-to-end AUV local motion planning method based on deep reinforcement learning. J. Marine Sci. Eng. **11**(9), 1796 (2023)
13. Xing, T., Wang, X., Ding, K., Ni, K., Zhou, Q.: Improved artificial potential field algorithm assisted by multisource data for AUV path planning. Sensors **23**(15), 6680 (2023)
14. Fujimoto, S., Van Hoof, H., Meger, D.: Addressing function approximation error in actor-critic method (2018)
15. Sarkar, T., Kalita, S.: A weighted critic update approach to multi agent twin delayed deep deterministic algorithm. In: 2021 IEEE 18th India Council International Conference (INDICON), pp. 1–6 (2021)

16. Fan, D., Shen, H., Dong, L.: Twin attentive deep reinforcement learning for multi-agent defensive convoy. Int. J. Mach. Learn. Cyber. **14**, 2239–2250 (2023)
17. Jia, J., Xing, X., Chang, D.E.: GRU-attention based TD3 network for mobile robot navigation. In: 2022 22nd International Conference on Control, Automation and Systems (ICCAS), pp. 1642–1647 (2022)
18. Chen, J.X., Jiang, D.M., Zhang, Y.N.: A hierarchical bidirectional GRU model with attention for EEG-based emotion classification. IEEE Access **7**, 118530–118540 (2019)
19. Author, F., Author, S.: Title of a proceedings paper. In: Editor, F., Editor, S. (eds.) CONFERENCE 2016. LNCS, vol. 9999, pp. 1–13. Springer, Heidelberg (2016)
20. Author, F., Author, S., Author, T.: Book title, 2nd edn. Publisher, Location (1999)
21. Author, F.: Contribution title. In: 9th International Proceedings on Proceedings, pp. 1–2. Publisher, Location (2010)

Wireless and Ubiquitous Networking

Active Optimization Routing Protocol Based on SDVN

Yufeng Chen[1], Xu Zhou[2], Xianfeng Weng[3(✉)], Bo Li[2], Zhengtao Xiang[2(✉)], Li Tian[2], and Minghao Yue[2]

[1] Institute of Automotive Engineers, Hubei University of Automotive Technology, Shiyan, China
chenyf_dy@huat.edu.cn
[2] School of Electrical and Information Engineering, Hubei University of Automotive Technology, Shiyan, China
xztcyf@163.com
[3] Noncommissioned Officer Academy of PAP, Hangzhou, China
wxfeng889@sina.com

Abstract. In the Internet of Vehicles (IoV) environment, the frequent changes in network topology are caused by the high-speed movement of vehicles, the diversity of road layouts, and the limited transmission range of vehicles. Therefore, selecting appropriate routing to ensure robust communication between vehicles is one of the main challenges in IoV. This paper focuses on the research of routing protocols in urban IoV scenarios. By integrating the IoV and software-defined network (SDN) architectures, a software-defined vehicular network architecture is constructed to enhance the reliability and efficiency of information transmission between vehicles. Utilizing the global view of the central controller, this study applies more vehicular information, including link stability, node congestion degree, and packet congestion degree, in route selection to improve path performance. Additionally, to address various complex scenarios in IoV, this paper proposes different deployment strategies, including proactive strategy, proactive-optimization strategy, and proactive-restriction strategy, to flexibly cope with complex scenarios and enhance the quality of communication services. Simulation results indicate that compared to the original passive strategy, the proactive-restriction strategy reduces end-to-end delay by 19%, increases delivery rate by 14.1%, and decreases the number of control messages by 39.8%.

Keywords: SDVN · Multipath routing · routing distribution strategy · Effect factor

1 Introduction

Intelligent Transportation Systems (ITS) are key technologies for solving traffic congestion issues and enhancing road safety. One of its core components, Vehic-

Supported by Open Experiment Project of CENI (China Environment for Network Innovations), China Major Scientific Infrastructure (No. 2022A0213).

X. Zhou et al. (Eds.): GPC 2024, LNCS 15225, pp. 169–186, 2026.
https://doi.org/10.1007/978-981-95-1346-8_11

ular Ad-hoc Networks (VANETs), utilizes vehicles as mobile nodes to establish an ad-hoc network between vehicles (Vehicle-to-Vehicle, V2V) and between vehicles and infrastructure (Vehicle-to-Infrastructure, V2I). The goal of VANET is to achieve real-time information exchange, and its characteristics include dynamic network topology, high vehicle mobility, and a distributed network structure, which significantly impact the efficiency of routing protocols [1] (Fig. 1).

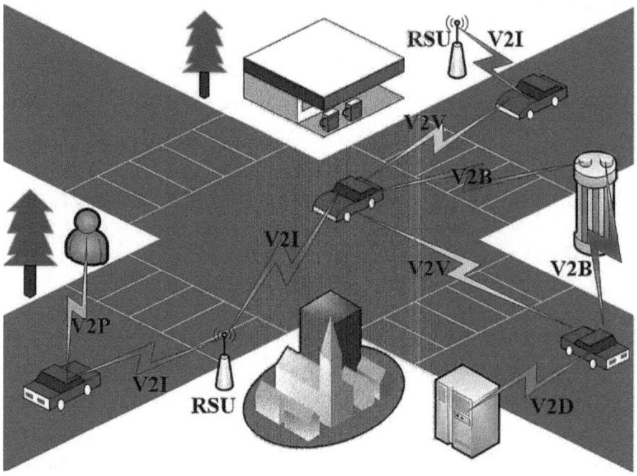

Fig. 1. V2X in the Internet of Vehicles.

In VANETs, the efficiency and stability of routing are core concerns. The complexity of vehicular environments and varying road conditions make the design of effective routing mechanisms a significant challenge [2]. Due to the critical importance of vehicle safety, the communication requirements in VANETs far exceed those of general data transmission. In emergency situations such as vehicle collisions or road obstructions, messages must be propagated in an extremely short time to enable vehicles to respond quickly. Therefore, routing protocols designed for VANETs must not only consider the efficiency of data transmission but also ensure its real-time nature and reliability. This requires routing protocols to function stably even in highly dynamic network environments.

Researchers have proposed using multipath routing to address these challenges. The study of multipath routing protocols in VANETs aims to improve the robustness and transmission efficiency of the network. Yuan et al. [3] extended AODV to discover multiple disjoint paths during the route discovery process, reducing the risk of single-path failures. This protocol adds additional fields such as hop count and path list in the control messages of AODV to utilize the existing routing information as much as possible. However, AOMDV still faces the issue of route cut-off, where it cannot find two completely separate paths if there are common intermediate nodes in the discovered paths. By introducing additional routing packets, AOMDV reduces route discovery latency, minimizes routing overhead, and achieves improvements in packet loss rate, route discovery frequency, and end-to-end delay.

As the VANET environment becomes increasingly complex, traditional multipath routing protocols face new challenges. Recent research has introduced SDN technology to enhance network flexibility and programmability by separating the control plane from the data plane [4]. Tasks are generated by the controller and transmitted to various network devices, thereby enabling the network to be flexible and programmable, allowing the system to perceive and respond to changing conditions and requirements [5].

Khan et al. [6] proposed a VANET hierarchical architecture based on 5G technology, integrating the centralized features of SDN and the 5G-supported Cloud-Radio Access Network (C-RAN), which can effectively allocate resources. This architecture significantly reduces the number of handovers between roadside units and network edge vehicles by employing clustering techniques.

Darabkh et al. [7] proposed the Innovative Cluster-Based Dual-Phase Routing Protocol (ICDRP-F-SDVN) based on fog computing and the SDN architecture, utilizing the latest technologies to overcome the shortcomings of traditional VANET routing. This protocol combines fog computing and the SDN architecture to provide a powerful new architecture that meets new requirements and addresses the challenges of the rapid growth of intelligent vehicles. Additionally, the ICDRP-F-SDVN protocol proposes a new clustering algorithm aimed at reducing remote communication within each cluster, and an effective control overhead reduction mechanism, with simulation results validating the effectiveness of these strategies.

Overall, with the development of VANET technology and the increasing complexity of the environment, the design and optimization of routing mechanisms have become more critical to meet the needs of intelligent transportation systems and improve the efficiency and reliability of communication. Although SDN brings new opportunities for VANET routing research, there are still some deficiencies. Traditional routing evaluation criteria often focus more on stability while lacking flexibility in dealing with complex topological changes. To address this issue, this paper proposes a new path effect factor. This factor considers the remaining bandwidth between links, node congestion levels, link duration, and the Euclidean distance to the destination node to avoid loops. Additionally, the paper proposes proactive, proactive-optimization, and proactive-restriction strategies to better adapt to complex topological changes.

The structure of this paper is as follows: Sect. 2 introduces the Software Defined Vehicular Network model; Sect. 3 details the data packet format definition, basic communication process, path effect factor, and various deployment strategies; Sect. 4 analyzes the simulation experiments and their results; finally, Sect. 5 summarizes the research findings and discusses future research directions.

2 Model

2.1 SDVN Model

The SDN architecture is an innovative network architecture proposed in recent years to address the shortcomings of traditional network architectures. In traditional network architectures, the functions responsible for data forwarding and control decisions are tightly integrated, leading to complex and rigid network configuration and management. This integration makes it difficult for the network's functions and performance to adapt to rapidly changing demands. Additionally, because new network protocols and applications often require hardware upgrades or replacements, the traditional integrated network architecture limits innovation. The SDN architecture aims to solve these issues by separating the control plane, which is responsible for network decision-making, from the data plane, which handles traffic forwarding [8]. This separation not only makes network control more flexible but also significantly simplifies network configuration and management.

The SDVN introduces the core concept of SDN, which is the separation of the control plane from the data plane, into vehicular ad-hoc networks, enabling flexible control and management of vehicular networks. In the SDVN architecture, a global controller is responsible for the control logic of the entire network. It guides the behavior of network devices by issuing policies and rules, thereby achieving dynamic adjustment and optimization of network traffic.

This paper adopts a logical SDN architecture model integrated with VANET, featuring a three-layer structure comprising the application plane, control plane, and data plane, as illustrated in Fig. 2. These layers collectively provide flexibility and centralized management for the network:

Data Plane: In the SDVN architecture, the data plane includes all vehicles and RSUs (Roadside Units). The data plane primarily collects data from each vehicle, such as location, speed, and direction, relying on the global positioning system. The controller can use this data to generate routing paths, determining the optimal route for data packets to reach their destination. Each vehicle sends data to the RSU through Hello messages, including information on direction, location, speed, vehicle ID, and node congestion levels. This data is stored in the RSU and forwarded to the global controller, enabling the global controller to have a comprehensive view.

Control Plane: The global controller obtains global vehicle and RSU data information. The control layer is responsible for intelligent decision-making and management of the network, including monitoring global vehicle status, proactively issuing routes, monitoring path status, calculating link costs, and generating optimal paths.

Application Plane: The application plane mainly consists of applications running on the network, such as the routing manager, link module, and proactive replacement module. These applications update routing strategies through the controller to ensure the real-time validity of routing information.

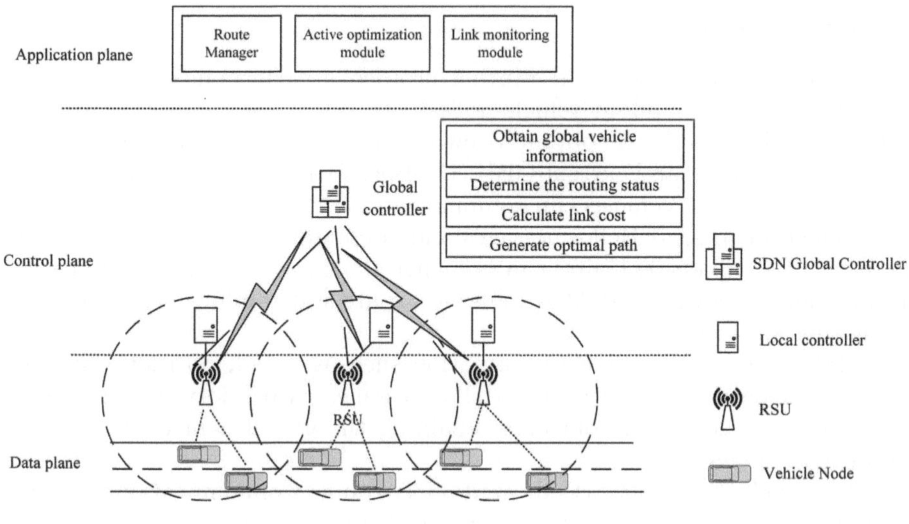

Fig. 2. SDVN architecture.

Applying the SDN architecture to the vehicular network environment can effectively meet diverse application requirements and balance the uncertainty of determining the optimal routing path with the complexity of discovering multiple disjoint paths from the source node to the destination node.

3 Proactive Optimization Strategy

In the communication process, several key steps are typically involved, such as packet definition (see Sect. 3.1), neighbor discovery (see Sect. 3.2), route requests (see Sect. 3.3), and sending communication packets. Particularly in the route request phase, this study adopts innovative methods to enhance routing efficiency and stability in vehicular networks. Specifically, when generating paths, the SDN global controller introduces new evaluation criteria, including considering the remaining bandwidth of links, node congestion levels, and link duration. Based on these factors, we define the "Path Effect Factor" (see Sect. 3.4), and use this factor along with the distance to the destination vehicle node to calculate the cost between links, thereby generating the minimum cost path.

Considering the rapid changes in topology in the vehicular network environment, the SDN global controller must also consider different route deployment strategies after generating the path. To address this, this study proposes various strategies, including the proactive strategy, proactive-optimization strategy, and proactive-restriction strategy (see Sect. 3.5). These strategies not only proactively optimize the current path to adapt to the dynamically changing network environment but also impose proactive restrictions on poorly performing paths to avoid inefficient routing choices.

3.1 Packet Design

To facilitate the analysis of the routing protocol's performance, the following assumptions are made about vehicle nodes:

1. Each vehicle node can obtain its own position via GPS and can acquire its own speed, direction, and current OBU (On-Board Unit) network status through the in-car controller. Vehicles can communicate with each other.

2. Vehicles can send Hello messages and communication information. Vehicles and RSUs (Roadside Units) can exchange Hello messages and route request information. RSUs will send ACK messages, route status information, and routing information to vehicles.

3. RSUs can obtain vehicle information such as speed, position, direction, and node congestion level through Hello messages. RSUs forward this vehicle information to the global controller, enabling the global controller to have a comprehensive view.

4. All vehicles use DSRC (Dedicated Short-Range Communications) technology. Hello messages are sent through service channels, while communication and routing control messages use control channels.

5. Vehicles discover neighbors through broadcasting.

Based on these assumptions, the design of the Hello data packet is as shown in Table 1. It includes the vehicle node's ID, position, speed, packet sequence number, heading, node congestion level, and remaining bandwidth, where the node congestion level represents the vehicle node's network status.

Table 1. Hello Data Packet Structure

Parameter	Meaning
SenderAddress	Vehicle ID
SenderPos	Vehicle position
SenderSpeed	Vehicle speed
Seq.No	Packet sequence number
Heading	Heading (vehicle angle)
k	Node congestion level
rwb	Node remaining bandwidth
cb	Node consumed bandwidth
Send.num	Node remaining data packets

The route request data packet, primarily includes the sender vehicle ID and the destination vehicle ID. The global controller generates routes based on the global view using these vehicle IDs and then issues them.

The routing data packet, as shown in Table 2, mainly includes the path and the destination node. The RSU issues the route to the source node, which stores the route in its routing table.

Table 2. Routing Data Packet Structure

Parameter	Meaning
FirstRoute	Primary path
SecondRoute	Backup path
Destination	Destination node

3.2 Neighbor Discovery

Neighbor discovery is a fundamental and critical function in various network protocols and applications, including mobile ad hoc networks and vehicular ad hoc networks. Its primary role is to enable network nodes to identify and record other nodes they can directly communicate with, i.e., their "neighbors," or communicate with edge facilities. Through neighbor discovery, vehicle nodes can exchange information with surrounding vehicles or facilities. In this study, when a vehicle broadcasts a Hello data packet, different actions are taken based on the type of receiver.

When the receiver is a neighboring vehicle, the neighboring vehicle will add or update its neighbor table. When the receiver is an RSU, the RSU will reply with an ACK message to notify the vehicle node that the Hello message has been received. The RSU will then forward the vehicle information to the SDN global controller, providing the SDN global controller with a global view. The neighbor discovery process is illustrated by the following pseudocode.

It is important to note that the vehicle information in Hello data packets is time-sensitive. Therefore, vehicles and the SDN global controller will periodically ($2.5 * \sigma$) delete expired vehicle information from the neighbor table, where σ is the interval for vehicles to send Hello data packets, set to 1 s in this study.

3.3 Communication Process

In the Internet of Vehicles (IoV), vehicle nodes exchange vehicle information through communication. The communication process is illustrated in Fig. 3. Vehicle nodes broadcast Hello messages to acquire neighbor information, which is then stored in the neighbor table. When an RSU receives a vehicle's Hello message, it forwards the vehicle information to the global controller, providing the global controller with a comprehensive view. The RSU sends an ACK message to inform the vehicle node whether it is within the RSU's communication range.

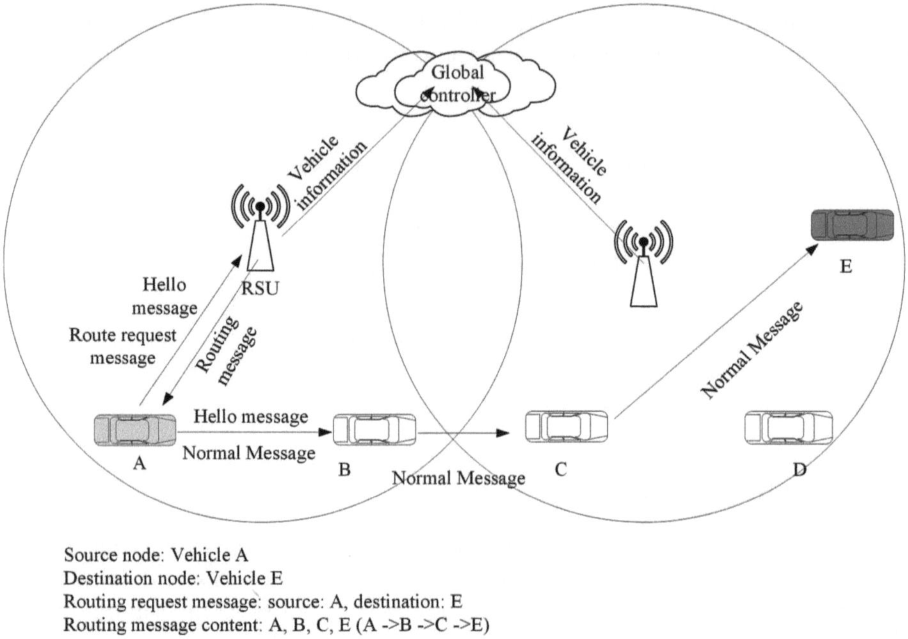

Fig. 3. Communication Process Diagram.

When there is a communication need between the source vehicle node and the destination vehicle node, the routing table is first checked for the existence of a route. If the route does not exist or is unavailable, the source vehicle node sends a route request to the RSU. Upon receiving the route request, the RSU forwards it to the global controller. The global controller, based on global vehicle information and various parameters, evaluates the performance of the global links, generates the optimal path, and issues it. If a route exists and is available, the communication data packet is sent directly. The structure of the routing table is shown in Table 3.

Table 3. Routing Table Structure

Parameter	Meaning
Destination	Destination vehicle ID
FirstRoute	Primary path
FirstRouteState	Primary path status
SecondRoute	Backup path
SecondRouteState	Backup path status

When vehicles move rapidly and the topology changes frequently, the global network state can be poor. In response to route requests, the global controller may be unable to generate an available route, or it may detect that the source node's path information is invalid. Therefore, the global controller will issue route status information to inform the source node about the path status, preventing the source node from using invalid paths to send data packets.

Considering the rapid changes in topology, this paper proposes different deployment strategies for routing issuance, as detailed in Sect. 3.5.

3.4 Path Effect Factor

In the SDVN architecture, the global controller obtains global vehicle information through the data plane, allowing for the selection of more comprehensive vehicle information. This study selects residual bandwidth, node congestion level, and link duration as reference indicators.

Residual Bandwidth: Similar to [9], the calculation method for residual bandwidth RBW_i is as follows:

$$RBW_i = \frac{C_{rate} - \delta}{C_{rate}} \tag{1}$$

The formula calculates the remaining bandwidth, where C_{rate} represents the upper limit of channel data rate, and δ represents the generation rate within the carrier range of vehicle , including MAC layer overhead, acknowledgment headers, and other overheads. Considering traffic variations and signal propagation characteristics, a weighted average is used to calculate δ .

$$\delta_t^- = (1-\alpha) \cdot \delta_{t-1}^- + \alpha \cdot \delta_t \tag{2}$$

with the formula including δ_t^-, which represents the data generation rate. It takes into account the data generation rates at the previous and current moments, with a focus on precision and agility. The value of α is set at 0.4.

Node Congestion Level: According to [10], the node congestion level κ for each vehicle is calculated as follows:

$$\kappa(v_i) = \frac{\tau(v_i)}{\zeta(v_i)} \tag{3}$$

where $\kappa(v_i)$ represents the node congestion level,$\tau(v_i)$ is the input flow rate, and $\zeta(v_i)$ is the service flow rate. The input flow rate refers to the number of packets that need to be forwarded by the vehicle within a certain time, and the service flow rate refers to the number of packets sent to the channel. Since the number of interface queues plays an important role in flow load variation, the vehicle node congestion information is collected.

Link Stability: Link duration value (LDV) is a key indicator of link stability in vehicular networks. LDV represents the duration of link existence [11], reflecting the stability of the link, as follows:

$$LDV(v_f, v_i) = \frac{R \pm Distance(v_f, v_i)}{\sqrt{(s_{v_f}\cos\theta_{v_f} - s_{v_i}\cos\theta_{v_i})^2 + (s_{v_f}\sin\theta_{v_f} - s_{v_i}\sin\theta_{v_i})^2}} \tag{4}$$

where R represents the communication distance between vehicles, v_f represents the current vehicle, and v_i represents the neighboring vehicle. When the two vehicles are moving in the same direction and v_f is behind v_i, the relative distance is $R - Distance(v_f, v_i)$. otherwise, it is $R + Distance(v_f, v_i)$. When the two vehicles are moving in opposite directions, if v_f is in front, the relative distance is $R - Distance(v_f, v_i)$. otherwise, it is $R + Distance(v_f, v_i)$. s_{v_f} represents the speed of the current vehicle, and s_{v_i} represents the speed of the neighboring vehicle. The denominator in the formula represents the relative speed between the two vehicles.

In this study, the link stability (LS) is set to a range of (0,1), where 1 represents a link that exists throughout the entire simulation period. However, when the LDV value exceeds the simulation time threshold δ, its exceeding portion becomes meaningless. Therefore, when the LDV value exceeds δ, the link stability is set to 1, indicating that the link exists throughout the simulation period. In the experiment, δ is set to 300.

$$LS = \begin{cases} 1 & LDV > \delta \\ \frac{LDV}{\delta} & LDV \leq \delta \end{cases} \tag{5}$$

Path Utility Factor (PUF): The path utility factor considers parameters such as residual bandwidth, node congestion level, and link stability to generate the path utility factor, as shown in the following formula:

$$PUF(v_f, v_i) = (\alpha(1 - RBW_i) + \alpha * \kappa_i + (1 - 2\alpha)(1 - LS(v_f, v_i))) \tag{6}$$

where $PUF(v_f, v_i)$ represents the path utility factor between the self vehicle and the neighboring vehicle, RBW_i represents the neighboring vehicle's residual bandwidth, κ_i represents the node congestion level, and $LS(v_f, v_i)$ represents the link stability between the self vehicle and the neighboring vehicle. In addition, this study considers the residual bandwidth of the link and the congestion level of the nodes, both expressed as percentages and adjusted by the weight parameter α. The selection of this weight parameter reflects the relative importance of these factors in the overall link quality evaluation, set to 0.25.

Path Generation: When selecting the transmission path, distance metrics are considered, prioritizing nodes closer to the target node under equal conditions. Specifically, the Euclidean distance between two points is obtained using the distance function $distance(v_i, v_d)$, where v_i and v_d represent the neighboring vehicle node and the target vehicle node, respectively. By dividing this distance by the communication distance, the theoretical minimum hop count to the target node can be estimated.

Considering that the value range of PUF is (0,1), while the value range of $distance$ is broader, an adjustment factor is applied to balance its impact in the overall cost calculation. The strategy multiplies the distance by an adjustment factor to achieve a weight ratio of 1:1 in cost evaluation, ensuring equal importance in path selection decision-making. The cost formula is shown as follows:

$$\cos t_{v_f}(v_f, v_i) = Distance(v_i, v_d)/R + PUF(v_f, v_i) \tag{7}$$

where $PUF(v_f, v_i)$ represents the path utility factor between the self vehicle and the neighboring vehicle, $Distance(v_i, v_d)$ represents the distance between the target vehicle and the neighboring vehicle, and R represents the vehicle communication distance. v_f represents the source vehicle, v_i represents the neighboring vehicle, and v_d represents the destination vehicle.

When the global controller obtains global vehicle information, it can generate a global cost table using the cost formula. The global minimum cost path is generated using the Dijkstra algorithm, with the path cost calculation formula as follows:

$$Cost_{p_m} = \sum_{v_i, v_d \in p_m}^{n} \cos t_{v_f}(v_i, v_d) \tag{8}$$

where $Cost_{p_m}$ represents the path cost, and $cost_{v_f}(v_i, v_d)$ is the cost of the link between intermediate nodes and neighboring nodes in the path. The path cost is the sum of the costs of adjacent nodes in the link, and the path with the minimum cost among all feasible paths is selected as the optimal path for issuance.

In the study of vehicular networks, the dynamic changes in network topology are an important consideration. The movement of vehicles continuously alters the network structure, necessitating real-time computation of the shortest path to optimize routing decisions. This study employs the Dijkstra algorithm to generate the minimum cost path. This algorithm not only accurately calculates the shortest path between nodes but also quickly updates path information when the network state changes.

3.5 Routing Deployment Strategies

Considering the high mobility and rapid topology changes in vehicular network environments, this paper proposes various innovative routing strategies aimed at optimizing the limitation in traditional multipath routing protocols, which only reinitiate route requests after both the primary and backup paths fail. These strategies include Proactive Strategy, Proactive-Optimization Strategy, and Proactive-Constraint Strategy, which collectively address dynamic changes in vehicular networks, enhancing routing efficiency and network stability. The specific strategies are as follows.

Passive Strategy (Original Strategy): in multipath routing, when the primary path fails, the feasibility of the backup path is assessed. If the backup path is available, it is used directly. When both the primary and backup paths fail, a new route request is initiated. It is important to note that the routing status information is issued by the RSU under the control of the global controller.

Proactive Strategy: Under this strategy, when the SDN global controller detects that both paths of the source vehicle node have failed, it immediately generates the minimum cost path and transmits it to the RSU. The RSU then issues the new route to the source vehicle node, thereby avoiding prolonged communication interruptions and reducing repeated route requests from the source node.

Proactive-Optimization Strategy: This strategy further optimizes the proactive strategy. Due to the frequent changes in road topology, the currently used path is often not the optimal one. To address this scenario, the Proactive-Optimization Strategy involves the global controller not only generating and issuing new routes but also continuously monitoring the shortest cost path for the source vehicle node. The SDN global controller compares the currently used path with the shortest cost path detected in real-time. If the cost of the new path is significantly lower than the current path, reaching a certain threshold as in Eq. (9), it forwards it to the RSU for route update. This ensures that the source vehicle node always uses the optimal path. Since the SDN global controller monitors the optimal path in real-time, multipath routing is converted to single-path routing.

$$\cos t_{p_{use}} > \varphi \cos t_{p_{now}} \tag{9}$$

where $\cos t_{p_{use}}$ is the cost of the current route, $\cos t_{p_{new}}$ is the cost of the new minimum route, and φ is the optimization factor, which is set to 1.2 in this study.

Proactive-Constraint Strategy: This strategy adds path generation constraints on top of the Proactive-Optimization Strategy. Considering that link failures often occur under poor link conditions, this strategy sets a path cost threshold for newly generated minimum cost paths. If the cost of the new path exceeds the set limit, the SDN controller will delay forwarding the path to reduce network load, wait for a better path to appear, and issue the source vehicle node's routing status to the RSU, which then forwards it to the source vehicle node.

By implementing these three strategies within the SDVN architecture, the routing response speed in vehicular networks is improved, and the network's stability and communication quality in highly dynamic environments are enhanced.

4 Result and Analysis

4.1 Simulation Settings

To ensure the accuracy and comprehensiveness of the VANET (Vehicular Adhoc Networks) scenario analysis, this study employs various advanced simulation tools to construct and test network models. Firstly, the network simulation platform OMNeT++ [12] is utilized to create complex network models and evaluate their performance. OMNeT++ supports a wide range of network protocols and device models, making it suitable for simulating VANET communication protocols and network topologies.

Furthermore, the SUMO (Simulation of Urban Mobility) traffic simulator [13] is integrated, which is specialized in simulating urban traffic environments.

SUMO can generate traffic flows on real city maps, reflecting actual vehicle behaviors and traffic patterns, and is used to test VANET performance and stability.

Finally, the Veins framework is adopted to integrate network and traffic simulations [14]. Veins combines OMNeT++ and SUMO, simulating vehicle mobility and wireless communication networks. This method evaluates VANET performance under different traffic and network conditions, providing a powerful tool for in-depth analysis of VANET performance and challenges.

The simulation experiments were set for 300 s, with the number of vehicles randomly generated as 100, 150, 200, 300, 400, and 500 within 0–100 seconds, and moving randomly on the road. Initial nodes appear randomly on the map, with minimum and maximum speeds set to increase topology. The communication radius is 350 m, the number of RSUs is 20, and the simulation experiments are run 100 times to obtain the average value. Specific parameters are shown in Table 4.

Table 4. Simulation Parameters

Simulation Parameters	Values
Vehicle Density	100, 150, 200, 300, 400, 500
Network Simulator	OMNeT++
Traffic Simulation	SUMO
Area	4000 m × 4000 m
Communication Protocol	IEEE 802.11p
Fading Model	Nakagami-m
Communication Radius	350 m
Speed	10–30 m/s
Number of RSUs	20
Simulation Time	300 s
Number of Simulations	100
Number of Connections	10

4.2 Simulation Scenario Settings

In the context of Intelligent Transportation Systems, exploring the dynamic characteristics of network topology is of paramount importance. This study aims to analyze the performance of Vehicular Ad-hoc Networks within typical urban simulation scenarios.

The simulation environment is constructed using real city map data provided by OpenStreetMap, accurately reflecting the actual urban road network layout and traffic characteristics. The selected urban scenario, as shown in Fig. 4, provides comprehensive background information for the simulation, thereby ensuring the practical applicability and relevance of the experimental results.

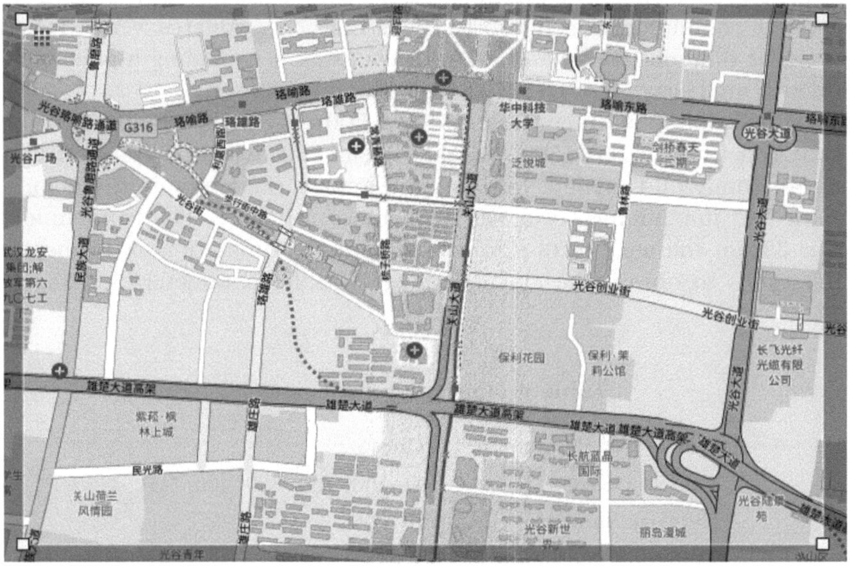

Fig. 4. Simulation Scenario

4.3 Evaluation Parameters

Quality of Service (QoS) parameter [15], including average end-to-end delay, delivery rate, and control message count, are used to measure the performance differences between various strategies. The descriptions are as follows:

Average End-to-End Delay: This is a crucial indicator for assessing network performance. It refers to the average total time taken for a data packet to travel from the sender to the receiver. The formula is as follows:

$$\frac{1}{m} \cdot \frac{\sum_{i=1}^{n} d_i}{\sum_{j=1}^{m} SPkt_j} \tag{10}$$

where d_i represents the delay in receiving packets, $SPkt_j$ represents number of packets sent by each source node is the number of packets sent by each source node, and m is the number of source nodes.

Delivery Rate: This is the ratio of the number of data packets received by the destination vehicles to the number sent by the source vehicles. It is obtained by comparing the number of packets sent by target vehicles to the number received. The formula is as follows:

$$\frac{1}{n}\sum_{i=1}^{n}\frac{RPkt_i}{SPkt_i} \tag{11}$$

where n represents the number of source and target nodes, $RPkt_i$ and $SPkt_i$ respectively represent the number of packets received by the target node and the number of packets sent by the source node.

Control Messages: This refers to the total number of routing control messages received or sent by source nodes. Routing control messages include route request messages, path status messages, and routing messages. The formula is as follows:

$$\sum_{i=1}^{n} cpk_i \tag{12}$$

where cpk_i is the number of control messages.

4.4 Analysis of Simulation Results

Figure 5 shows the end-to-end delay under the simulated scenario. The results indicate that the active-limitation strategy maintains a favorable delay, as it restricts the transmission of poor-performance packets, thereby alleviating the overall network status. As illustrated in Fig. 5, the active-limitation strategy reduced the delay by 19% compared to the passive strategy and by 20.5% compared to the active strategy.

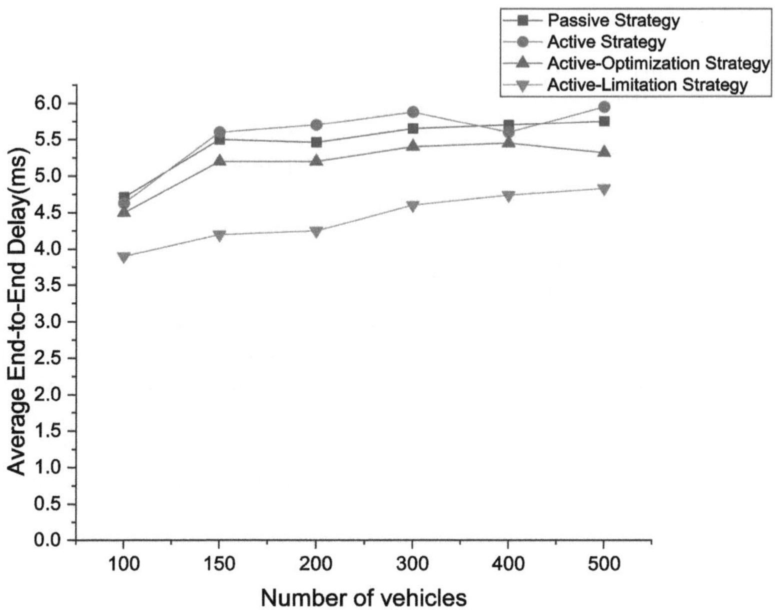

Fig. 5. End-to-End Delay in Simulation Scenario.

Figure 6 displays the delivery rates of various strategies in the simulation scenario. The results demonstrate that in this context, the active-limitation strategy outperforms the other three strategies. This is attributed to the strategy's limitation on disseminating routes with poor performance, which alleviates network congestion and reduces the occurrence of route failures. Meanwhile, the SDN global controller continues to seek new paths for the source vehicle node. The experiment shows that the delivery rate of the active-limitation strategy improved by 14.1% compared to the passive strategy.

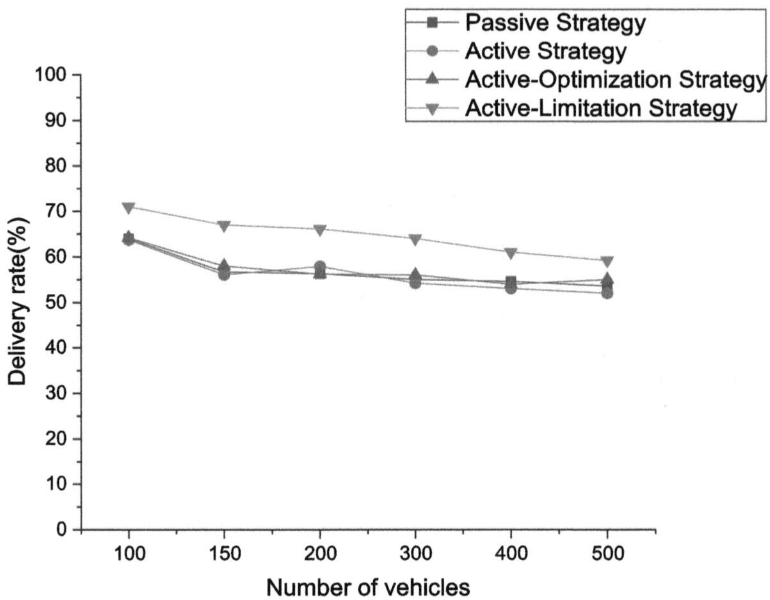

Fig. 6. Delivery Rate in Simulation Scenario.

Figure 7 shows the number of control messages for the four strategies in the simulated scenario. The results reveal that the active-limitation strategy performs optimally by actively restricting the dissemination of poor-performance routes, which decreases the occurrence of route failures and reduces the number of control messages sent. The active-limitation strategy reduced control messages by 39.8% compared to the passive strategy, by 35.5% compared to the active strategy, and by 7.4% compared to the active-optimization strategy. The active-optimization strategy reduced control messages by 35.1% compared to the passive strategy and by 30.4% compared to the active strategy.

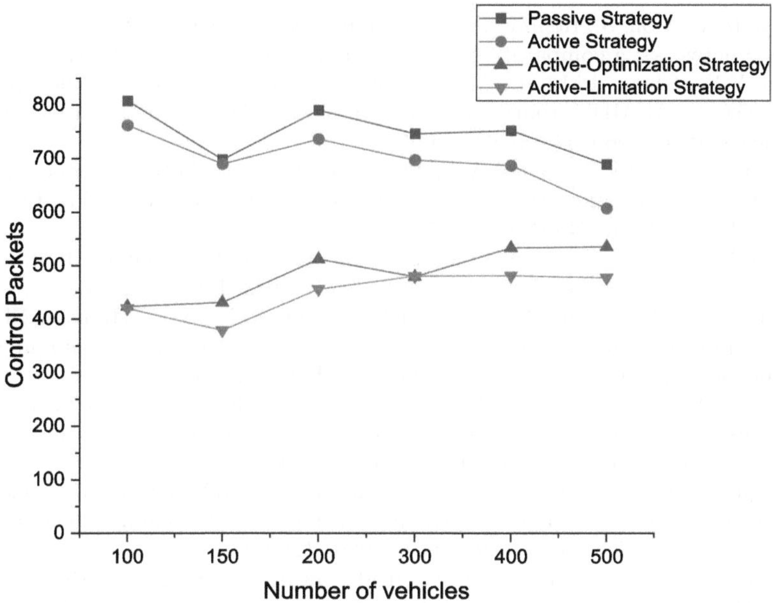

Fig. 7. Control Messages in Simulation Scenario.

5 Conclusion

This paper proposes a novel Path Effect Factor to generate the least-cost path, utilizing the global perspective of the SDN architecture to find more efficient routes. Building upon the original passive strategy, active, active-optimization, and active-limitation strategies are introduced. Simulation experiments conducted using OMNeT++, SUMO, and Veins demonstrate that both active-optimization and active-limitation strategies perform well in terms of delivery rate, end-to-end delay, and the number of control messages. The experiments validate the effectiveness of the active-limitation strategy in urban simulation scenarios, showing its superior performance in delivery rate, end-to-end delay, and control message count.

Future work may include: 1) Enhancing the performance of backup routes to reduce the likelihood of simultaneous failure with primary routes. 2) Exploring the integration of Software-Defined Vehicular Network environments with 5G to improve communication quality. 3) Considering different parameters to assess route performance and improve delivery rates in urban simulation scenarios.

References

1. Hussein, N.H., Yaw, C.T., Koh, S.P., et al.: A comprehensive survey on vehicular networking: communications, applications, challenges, and upcoming research directions. IEEE Access **10**, 86127–86180 (2022)

2. Nazib, R.A., Moh, S.: Reinforcement learning-based routing protocols for vehicular ad hoc networks: a comparative survey. IEEE Access **9**, 27552–27587 (2021)
3. Yuan, Y.H., Chen, H.M., Jia, M.: An optimized ad-hoc on-demand multipath distance vector (AOMDV) routing protocol. In: 2005 Asia-Pacific Conference on Communications, pp. 569–573. IEEE (2005)
4. Al-Shareeda, M.A., Alsadhan, A.A., Qasim, H.H., et al.: Software defined networking for internet of things: review, techniques, challenges, and future directions. Bull. Electr. Eng. Inform. **13**(1), 638–647 (2024)
5. Sohail, M., Latif, Z., Javed, S., et al.: Routing protocols in vehicular adhoc networks (VANETs): a comprehensive survey. Internet Things, 100837 (2023)
6. Khan, A.A., Abolhasan, M., Ni, W.: 5G next generation VANETs using SDN and fog computing framework. In: 2018 15th IEEE Annual Consumer Communications & Networking Conference (CCNC), pp. 1–6. IEEE (2018)
7. Darabkh, K.A., Alkhader, B.Z., Ala'F, K., et al.: ICDRP-F-SDVN: an innovative cluster-based dual-phase routing protocol using fog computing and software-defined vehicular network. Veh. Commun. **34**, 100453 (2022)
8. Siddiqui, S., Hameed, S., Shah, S.A., et al.: Toward software-defined networking-based IoT frameworks: a systematic literature review, taxonomy, open challenges and prospects. IEEE Access **10**, 70850–70901 (2022)
9. Alzamzami, O., Mahgoub, I.: Link utility aware geographic routing for urban VANETs using two-hop neighbor information. Ad Hoc Netw. **106**, 102213 (2020)
10. Elappila, M., Chinara, S., Parhi, D.R.: Survivable path routing in WSN for IoT applications. Pervasive Mob. Comput. **43**, 49–63 (2018)
11. Chahal, M., Harit, S.: Network selection and data dissemination in heterogeneous software-defined vehicular network. Comput. Netw. **161**, 32–44 (2019)
12. Obelovska, K., Danych, I.: Adoption of the OMNET++ simulator for the computer networks learning: a case study in CSMA schemes. In: The International Conference on Artificial Intelligence and Logistics Engineering, vol. 135, no. 4, pp. 234–243 (2022)
13. Krajzewicz, D., Erdmann, J., Behrisch, M., et al.: Recent development and applications of SUMO-simulation of urban mobility. Int. J. Adv. Syst. Meas. **5**(3&4), 128–153 (2012)
14. Sommer, C., et al.: Veins: the open source vehicular network simulation framework. In: Virdis, A., Kirsche, M. (eds.) Recent Advances in Network Simulation. EICC, pp. 215–252. Springer, Cham (2019). https://doi.org/10.1007/978-3-030-12842-5_6
15. Sharef, B., Alsaqour, R., Alawi, M., et al.: Robust and trust dynamic mobile gateway selection in heterogeneous VANET-UMTS network. Veh. Commun. **12**, 75–87 (2018)

Pervasive and Green Computing

OOB-CM: Enhancing OOB Estimate for Data Valuation via Curriculum Learning and Multi-round Voting

Yuqi Jiao, Guanglu Sun, Zhenhua Wang, Jing Jin, Fei Lang(✉), and Suxia Zhu

Harbin University of Science and Technology, Harbin, China
{2110402069,2320400018,2210403070}@stu.hrbust.edu.cn,
{sunguanglu,langfei,zhusuxia}@hrbust.edu.cn

Abstract. Data valuation is an emerging area in machine learning that aims to assess the contribution of training data to a model's performance on unseen data. Among existing approaches, Data-OOB leverages out-of-bag (OOB) estimate to enable efficient data valuation and has shown promising results in tasks such as noisy data detection. However, our analysis reveals two key limitations of Data-OOB: (1) the weak learners it employs are vulnerable to noisy data, and (2) it suffers from information loss during the exploitation of weak learners. To address these issues, we propose a novel data valuation framework called OOB-CM. This method assigns a dynamic threshold to each data point and uses it to define a personalized curriculum for every weak learner. The curriculum selectively filters training samples at each stage of training, shielding weak learners from the influence of noise. Furthermore, we introduce a multi-round voting strategy that aggregates the discriminative signals from weak learners by selecting the top-p predicted classes for each sample. This strategy significantly improves the utilization of weak learners' predictive power. We conduct comprehensive evaluations across multiple scenarios, and the results consistently demonstrate that OOB-CM outperforms existing methods in terms of robustness and effectiveness.

Keywords: Data valuation · Data quality · Machine learning · Curriculum learning

1 Introduction

Data point plays a central role in the success of machine learning models. However, not all data points are equally useful—some contribute more to model performance, while others may be noisy or even harmful. This observation has led to the development of data valuation methods, which aim to measure the importance of each sample for a given learning task. Accurate data valuation can improve data selection, enhance model robustness, and reduce training costs.

X. Zhou et al. (Eds.): GPC 2024, LNCS 15225, pp. 189–203, 2026.
https://doi.org/10.1007/978-981-95-1346-8_12

Standard approaches for estimating data value are often based on the Shapley value [16,24,26,30]. However, computing exact Shapley values involves evaluating marginal contributions [4,10] over all possible subsets, which leads to prohibitive computational costs. To address this, various lightweight and approximate methods have been proposed. Among these alternatives, Data-OOB [19] has recently attracted considerable attention as an efficient and practical solution. It leverages out-of-bag (OOB) estimate [5,6] from the bagging model to assess the importance of individual data points. Compared to other methods in the field, Data-OOB demonstrates strong performance in tasks such as noise detection [27], data point removal, and data selection [8,12,13]. However, our study reveals that Data-OOB suffers from two critical shortcomings: (1) Sensitivity to noisy data. Since Data-OOB does not require a separate validation set for utility evaluation, the bagging model of Data-OOB tends to overfit noisy samples when noise is present in the training data. In this condition, Data-OOB often assigns disproportionately high values to noisy data points—an outcome that is clearly undesirable and misaligned with the goal of accurate data valuation. (2) Low utilization of weak learners. In Data-OOB, weak learners primarily vote based on the predicted class with the highest confidence. Consequently, when a data point's true label differs from its highest-confidence prediction, the votes relevant to the true label are ignored. This leads to under-utilization of valuable information from weak learners and limits the method's ability to capture meaningful signals related to true label support.

In this paper, we first propose a robust training method for weak learners incorporating curriculum learning [2]. The approach employs a curriculum learning guided by a dynamic threshold to strategically regulate the learning of training data points, where the dynamic threshold is adaptively set based on the model's memorization effect [1]. The memorization effect suggests that models tend to learn simple (i.e., clean and consistent) samples early in training, followed by more difficult ones. Given the inter-class inconsistency and misleading nature of noisy data, such samples are typically treated as hard examples and learned at later stages. Inspired by this, we introduce a dynamic threshold for each data point within the bootstrapped subsets, which tracks the historical memorization strength of the weak learner over time. The curriculum learning strategy leverages this dynamic threshold to design a tailored curriculum for each weak learner, guiding them to selectively learn from the training samples. This approach effectively suppresses the adverse effects of noisy data on weak learners during the early stages of training. Furthermore, to enhance the information utilization of weak learners, we design a multi-round voting mechanism based on the top-p predicted classes. Inspired by runoff voting, this approach enables each weak learner to vote not only for its top-1 prediction but also to cast additional votes for a set of top-ranked classes. By aggregating votes across multiple rounds, the method captures richer evaluation signals from weak learners, especially when the true label does not align with the most confident prediction. This leads to more informed and reliable data value estimates.

The key contributions of our paper are as follows:

- We propose a robust training method for weak learners based on curriculum learning. This method dynamically determines adaptive thresholds to automatically filter out noisy data points, enhancing the resilience of weak learners against noise.
- We introduce a multi-round voting data valuation strategy. This strategy progressively aggregates the discriminative information from weak learners through iterative filtering, enabling a more precise and robust estimation of each data point's value.
- We conduct a systematic evaluation of OOB-CM on 11 public datasets. Our results show that OOB-CM outperforms existing methods.

2 Related Work

Data Valuation. Most existing data valuation methods are based on marginal contribution, which measures the change in model performance when a data point is removed from a specific subset. The simplest of these methods is Leave-One-Out (LOO) [17], which evaluates each data point's contribution by considering its removal from the entire training set. A more representative approach is Data Shapley, which aggregates marginal contributions over all possible subsets to comprehensively assess the value of each data point. As a result, Data Shapley generally outperforms LOO. Several extensions of Data Shapley have been proposed. For instance, Beta Shapley [18] and Data Banzhaf [28] relax the fairness axioms of Shapley values using semivalues, making the methods more flexible in practice. Other variants, such as CS Shapley [23] and P Shapley [31], modify the utility function of Data Shapley to provide a fairer evaluation of data contributions. However, accurately estimating Shapley values incurs significant computational overhead, which limits their applicability to large-scale datasets. To address this issue, Jia et al. [11] employ K-Nearest Neighbors models to replace the default utility function, while Lin et al. [20] use LASSO regression to approximate the Shapley values.

Another line of data valuation methods is based on out-of-bag (OOB) estimate, which have been shown to maintain order consistency with infinitesimal jackknife influence function [3,5]. Representative approaches in this category include Data-OOB and 2D-OOB [25]. However, these methods are susceptible to the negative impact of noisy data and struggle to fully exploit the discriminative information provided by weak learners. In addition, there exist several other notable data valuation methods that are based on the reinforcement learning [33], the influence function [17], the volume of the data matrix [32], and the

OT distance [15], among others. In this paper, we take Data-OOB as a representative baseline and propose a novel data valuation framework to address its limitations in terms of robustness to noise and utilization of discriminative information.

Curriculum Learning. Curriculum learning [2,29] is a training strategy inspired by the human learning process, where models are presented with easier examples first and gradually exposed to more difficult ones. Bengio et al. [2] first introduced this concept, showing that organizing training data in a meaningful order can improve convergence speed and generalization performance. In addition, curriculum learning is commonly applied in areas such as learning from noisy labels (LNL) [34] and reinforcement learning [21]. In this work, we incorporate curriculum learning into the training of weak learners to protect them from the adverse effects of noisy data.

3 Proposed Method: OOB-CM

In this section, we first review and analyze the classical data valuation framework, Data-OOB. Then, we propose a novel framework, OOB-CM, which enhances robustness to noise and improves the utilization of weak learners compared to Data-OOB.

3.1 Preliminaries

For $d, c \in \mathbb{N}$, the input space and the output space are defined as $\mathcal{X} \in \mathbb{R}^d$ and $\mathcal{Y} \in \mathbb{R}^c$. Consider a training dataset $\mathcal{D} = \{(x_i, y_i)\}_{i=1}^n$, where $x_i \in \mathcal{X}$ and $y_i \in \mathcal{Y}$ are the input and label of the i-th data point.

In the Data-OOB setting, the training dataset will be bootstrap-sampled B times. Let $b \in [B]$, the vector $w_b = (w_{b1}, ..., w_{bn})$ records the number of times each training sample is selected during the b-th round of bootstrap sampling. Based on w_b, the weak learner f_b is defined as: $f_b : \mathcal{X} \rightarrow \mathcal{Y}$, which represents a weighted risk minimization:

$$f_b := \text{argmin}_{f \in \mathcal{F}} \frac{1}{n} \sum_{i=1}^{n} w_{bi} \ell(y_i, f(x_i)), \tag{1}$$

where $\ell : \mathcal{Y} \times \mathcal{Y} \rightarrow \mathbb{R}$ is a loss function. For the training data point (x_i, y_i), the Data-OOB value is defined as:

$$\psi_{\text{oob}}(x_i, y_i) = \frac{\sum_{b=1}^{B} \mathbb{1}(w_{bi} = 0) G(y_i, f_b(x_i))}{\sum_{b=1}^{B} \mathbb{1}(w_{bi} = 0)}, \tag{2}$$

where $G(y_i, f(x_i))$ is a score function. In the classification settings, $G(y_i, f(x_i)) = \mathbb{1}(y_i = f_b(x_i))$. Therefore, Eq. (2) can be viewed as Data-OOB leveraging weak learners to perform an OOB estimate. It further reveals that the data point

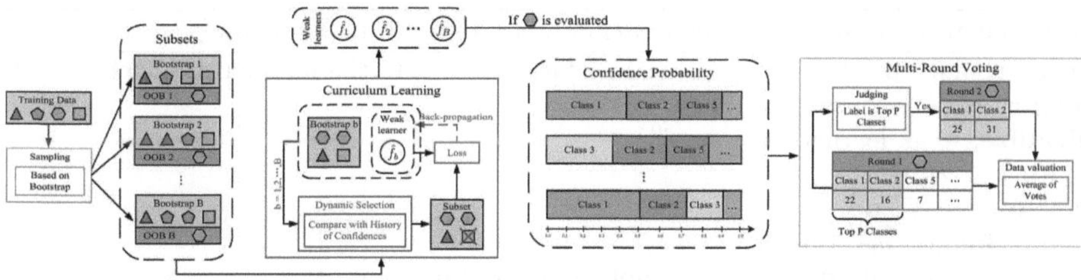

Fig. 1. The training dataset is partitioned into B bootstrap sets and their corresponding out-of-bag (OOB) sets via bootstrap sampling. For each bootstrap set, a weak learner is trained under a curriculum learning strategy. After training, the weak learner performs a voting-based evaluation on data points of OOB set. During the voting process, we adopt a multi-round voting strategy to progressively aggregate the votes from weak learners for data valuation. Detailed implementation can be found in Sect. 3.2.

(x_i, y_i) will receive a higher score when it is better recognized by the weak learners.

The above analysis suggests that the effectiveness of Data-OOB is contingent on the quality of the weak learners. When the training dataset includes noisy data points—such as mislabeled data points—the weak learners may overfit these instances. Under this condition, Data-OOB may incorrectly assign high values to these noisy data points.

Furthermore, the analysis reveals that the weak learner's vote for a data point is determined by its most confident prediction. This indicates that Data-OOB fails to capture the full spectrum of discriminative information provided by weak learners. As a result, valuable signals that could enhance data valuation are overlooked, leading to suboptimal performance in noisy or ambiguous scenarios.

3.2 OOB-CM

In this section, we propose a novel data valuation framework, OOB-CM. This framework addresses the issues identified in the previous section by combining curriculum learning with dynamic thresholds and a multi-round voting scheme based on the top-p prediction (Fig. 1).

As widely recognized, a model $f : \mathcal{X} \rightarrow \mathcal{Y}$ progressively increases its memorization strength of training data points during the learning process. In classification tasks, memorization strength for a data point (x, y) is typically represented by the model's confidence that x belongs to the ground-truth label y: $\mathrm{P}(y = f(x))$.

Since models memorize clean data before noisy examples, selecting data in early training stages can help avoid overfitting to noise. As illustrated in Algorithm 1, we leverage historical confidence to set a dynamic threshold. This threshold is used to design a curriculum that guides the training of the weak learner, thereby reducing the negative impact of noisy data.

Algorithm 1. Training Weak Learner via Curriculum Learning

Input: bootstrap set $S_1, \ldots, S_B$, scaling factor $\beta, \lambda > 0$, the number of epochs $T > 0$
Output: weak learners $\hat{f}_1, \ldots, \hat{f}_B$
Initialize: weak learners $f_1^0, \ldots, f_B^0$

```
1: for b ∈ [B] do
2:    for t ∈ [T] do
3:       for (x_i, y_i) ∈ S_b do
4:          Compute dynamic threshold τ_t^i using Equation (4)
5:       end for
6:       Update the weak learner f_b^t using Equation (3)
7:    end for
8:    Derive weak learner f̂_b = f_b^T
9: end for
```

Specifically, there are B bootstrap sets $S_1, ..., S_B$ are obtained through B rounds of bootstrap sampling. Assume that training a weak learner requires T epochs. For $b \in [B]$ and $t \in [T]$, f_b^{t-1} represents the weak learner that has been trained on the S_b and is about to start its t-th training epoch. At the t-th epoch, f_b^{t-1} is updated based on the following loss:

$$\ell_t = \sum_{(x_i, y_i) \in S_b} \mathbb{1}(\mathrm{P}_{t-1}^i > \tau_t^i)\ell(y_i, f_b^{t-1}(x_i)), \tag{3}$$

where $\mathrm{P}_{t-1}^i = \mathrm{P}(y_i = f_b^{t-1}(x_i))$, τ_t^i is a dynamic threshold of (x_i, y_i). In other words, each data point is assigned a dynamic threshold. If the confidence of (x_i, y_i) exceeds its threshold, we consider it trustworthy and include it in model training. For τ_t^i, it is updated using historical information:

$$\tau_t^i = \mu_t^i - \beta \cdot \sigma_t^i, \tag{4}$$

where β is the scaling factor. Assume that the label for each class is represented by $Y_1, ..., Y_c$, μ_t^i is the mean of historical confidence:

$$\mu_t^i = \frac{1}{t} \sum_{k=0}^{t-1} \max(\mathrm{P}_k^{i,1}, \ldots, \mathrm{P}_k^{i,c}). \tag{5}$$

For $j = 1, ..., c$, $\mathrm{P}_k^{i,j} = \mathrm{P}(Y_j = f_b^k(x_i))$. To simplify the computation, we apply a soft update to μ_t^i:

$$\mu_t^i = \lambda \cdot \mu_{t-1}^i + (1 - \lambda) \cdot \max(\mathrm{P}_{t-1}^{i,1}, \ldots, \mathrm{P}_{t-1}^{i,c}), \tag{6}$$

where λ is the scaling factor, μ_0^i is set to 0. In addition, σ_t^i is the standard deviation of the highest confidence scores in each epoch from the beginning up to epoch $t - 1$. Namely, we hope to track the historical standard deviation of the maximum confidence scores to prevent the threshold from becoming overly extreme.

Algorithm 2. Enhancing OOB Estimate for Data Valuation via Curriculum Learning and Multi-Round Voting

Input: training dataset $\mathcal{D} = \{(x_i, y_i)\}_{i=1}^n$, scaling factor $\beta, \lambda > 0$, the number of epochs $T > 0$, the number of weak learners $B > 0$, the number of classes $c > 0$
Output: data values $\psi(x_i, y_i), \ldots, \psi(x_n, y_n)$
Initialize: data values $\psi(x_i, y_i) \leftarrow 0$ for $i = 1, \ldots, n$

Sample subsets from $\mathcal{D}$ using bootstrap:

$$\{S_1, O_1\}, \ldots, \{S_B, O_B\} \leftarrow \text{Bootstrap Sample}(\mathcal{D})$$

Train weak learners $\hat{f}_1, \ldots, \hat{f}_B$ using Algorithm 1
for $i \in [n]$ **do**
  for $j_1 \in [c]$ **do**
    Record the class votes $N_{j_1}^{R1}(x_i, y_i)$ using Equation (7)
  end for
  Record the top-p classes c_p using Equation (8)
  if $0 \neq \sum_{j_2 \in c_p} \mathbb{1}(Y_{j_2} = y_i)$ **then**
    Update $\psi(x_i, y_i)$ using Equation (9)
  else
    for $j_2 \in c_p$ **do**
      Record the class votes $N_{j_2}^{R2}(x_i, y_i)$ using Equation (10)
    end for
    Update $\psi(x_i, y_i)$ using Equation (11)
  end if
end for

Next, we propose an evaluation strategy based on multiple rounds of voting. As shown in Algorithm 2, it first uses the discriminative information from weak learners to select the top-p candidate classes for each data point, and then aggregates additional discriminative signals based on these selected classes for final evaluation.

Specifically, we first record the class-wise voting results of the weak learner for each data point in the first round:

$$N_{j_1}^{R1}(x_i, y_i) = \sum_{b=1}^{B} \mathbb{1}((x_i, y_i) \notin S_b)\mathbb{1}(Y_{j_1} = \hat{f}_b(x_i)), \tag{7}$$

where $j_1 \in [c]$, $N_{j_1}^{R1}(x_i, y_i)$ is the voting results of all weak learners for (x_i, y_i) in the first round. Therefore, the top-p classes can be defined as:

$$\begin{aligned} c_1(x_i, y_i) &= \arg\max_{j_1 \in [c]} N_{j_1}^{R1}(x_i, y_i) \\ &\vdots \\ c_p(x_i, y_i) &= \arg\max_{j_1 \in c'} N_{j_1}^{R1}(x_i, y_i), \end{aligned} \tag{8}$$

Table 1. A summary of datasets used in our experiments.

Name	Sample size	Input dimension	Number of Classes
adult	48842	108	2
2dplanes	40768	10	2
jannis	57580	54	2
CPU	8192	21	2
creditcard	30000	23	2
digits	1797	64	10
pol	15000	48	2
fried	40768	10	2
nomao	34465	89	2
electricity	38474	6	2
covertype	581012	54	7

where $c' = [c]\backslash\{c_1(x_i, y_i), \ldots, c_{p-1}(x_i, y_i)\}$. If the truth class isn't included in the top-p candidates, the value of (x_i, y_i) is set to:

$$\psi(x_i, y_i) = \frac{\sum_{j_1 \in [c]} \mathbb{1}(Y_{j_1} = y_i) N_{j_1}^{R1}(x_i, y_i)}{\sum_{b=1}^{B} \mathbb{1}((x_i, y_i) \notin S_b)}. \tag{9}$$

Otherwise, we proceed to the second round of aggregated voting:

$$N_{j_2}^{R2}(x_i, y_i) = \sum_{b=1}^{B} \mathbb{1}((x_i, y_i) \notin S_b) \mathbb{1}(j_2 = \arg\max_{j \in c_p} \mathrm{P}(Y_j = \hat{f}_b(x_i))), \tag{10}$$

where $j_2 \in c_p$, $c_p = \{c_1(x_i, y_i), ..., c_p(x_i, y_i)\}$. Under this setting, the value of (x_i, y_i) is:

$$\psi(x_i, y_i) = \frac{\sum_{j_2 \in c_p} \mathbb{1}(Y_{j_2} = y_i) N_{j_2}^{R2}(x_i, y_i)}{\sum_{b=1}^{B} \mathbb{1}((x_i, y_i) \notin S_b)}. \tag{11}$$

The above process employs multiple rounds of voting to aggregate the discrimination information provided by weak learners for each data point. This approach significantly enhances the utilization of weak learner outputs and ensures a fairer and more reasonable data evaluation. In the following section, we demonstrate the effectiveness of the proposed method through a series of classic experiments.

4 Experiments

In this section, the performance of OOB-CM is comprehensively evaluated through experiments on detecting noisy data, removing data points, and learning with weighted data points. In addition, ablation studies are conducted on different components of OOB-CM to verify their effectiveness.

Table 2. F1-score of noisy data detection. The 6 data valuation methods are compared: AME, Beta Shapley (BS), Data-OOB (Data), Data Shapley (DS), LOO, and OOB-CM (ours). Bold numbers indicate the best method. In all cases, the proposed OOB-CM achieves the highest score.

Dataset	F1-score					
	AME	BS	Data	DS	LOO	ours
adult	0.275	0.543	0.593	0.543	0.295	**0.606**
2dplanes	0.329	0.598	0.665	0.591	0.313	**0.672**
creditcard	0.234	0.584	0.588	0.579	0.293	**0.593**
digits	0.388	0.492	0.643	0.498	0.306	**0.732**
pol	0.306	0.543	0.686	0.532	0.264	**0.701**
jannis	0.275	0.408	0.471	0.385	0.277	**0.481**
fried	0.301	0.552	0.646	0.533	0.273	**0.658**
nomao	0.346	0.656	0.754	0.674	0.254	**0.787**
CPU	0.250	0.721	0.777	0.719	0.287	**0.805**
electricity	0.227	0.469	0.529	0.460	0.235	**0.533**

Baseline Methods. OOB-CM is compared against several representative data valuation methods, including Data-OOB [19], AME [20], Beta Shapley [18], Data Shapley [9], and LOO [17]. All methods are integrated within the standardized framework 'OpenDataVal' [14], using its default parameter configurations. For OOB-CM, we configure the framework to perform two valuation rounds. The parameters p, λ, and β are empirically set to 2, 0.99, and 0.01, respectively. In our experiments, the logistic regression model serves as the weak learner in both Data-OOB and OOB-CM. Moreover, it is adopted as the utility evaluation model for Data Shapley, Beta Shapley, AME, and LOO to ensure consistency across methods.

Datasets. As shown in Table 1, a total of 11 datasets are used to evaluate the performance of OOB-CM. These datasets are obtained from either 'OpenML' [7], 'OpenDataVal' [14] or the Python package 'scikit-learn' [22]. Following prior work, each dataset is split into a training dataset, a validation dataset, and a test dataset. The training dataset is used for data valuation, the validation dataset is used to evaluate the utility function, and the test dataset is used to assess test accuracy. In accordance with the setting of Kwon et al. [19], the sizes of the training, validation, and test datasets are fixed at 1000, 100, and 3000, respectively. Since the 'digits' dataset is relatively small, its training, validation, and test datasets are set to 700, 70, and 1000 samples, respectively. In our experiments, 20% of the training instances are randomly selected and their labels are flipped to inject noise.

4.1 Noisy Data Detection

Due to measurement and annotation errors, noisy data points are often present in real-world datasets. Numerous studies have shown that such data points can

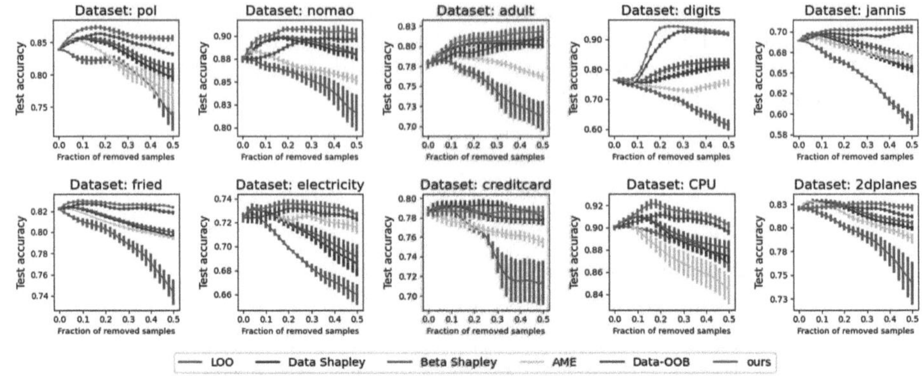

Fig. 2. Performance change curves of models after sequentially removing data points based on data valuation from low to high. Higher curves indicate better performance of the data valuation method. Error bars represent the 95% confidence interval of 50 independent experiments.

significantly harm model generalization. In this section, we investigate the ability of OOB-CM to identify noisy data points.

Table 2 shows the F1-score for noise detection using OOB-CM and other data valuation methods. Following Kwon et al., the training dataset is clustered into two groups using the K-means algorithm based on the data values. The cluster with the lower average value is treated as the noisy group, which is compared with the true noisy data points to calculate the F1-score. In summary, OOB-CM achieves the best performance, which suggests that it effectively assigns the lowest data values to noisy data points. Since OOB-CM leverages curriculum learning to effectively suppress the negative impact of noisy data points on weak learners during the early training phases, OOB-CM achieves better results than Data-OOB. In addition, Data Shapley and Beta Shapley benefit from more refined sampling procedures, so they perform better in noise detection than AME and LOO.

4.2 Point Removal Experiment

Since data values represent the contribution of data points to model performance, removing high-value or low-value data points is expected to decrease or increase the model's test performance, respectively. To evaluate whether OOB-CM can effectively distinguish between these two types of data points, we conduct a data removal experiment.

First, training data points are progressively removed in ascending order based on their data values. After each removal, the remaining data points are used to train a logistic regression model, which is then evaluated on the test dataset. As shown in Fig. 2, OOB-CM achieves the highest test performance, demonstrating its effectiveness in identifying low-value data points. Furthermore, Fig. 2 indicates that OOB-CM leads to the fastest improvement in model performance

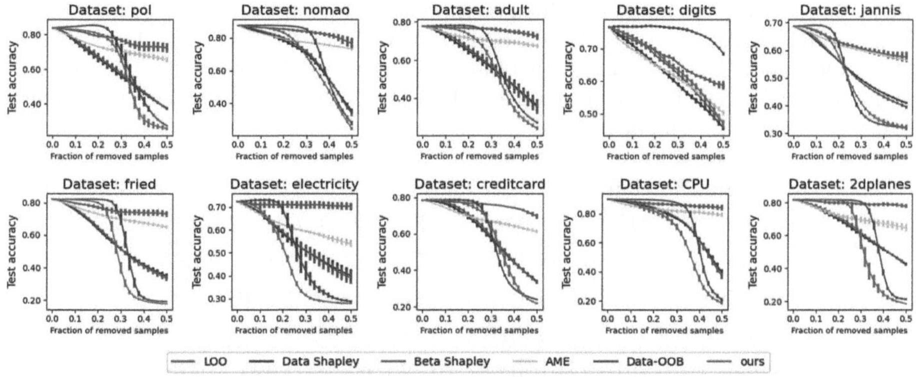

Fig. 3. Performance change curves of models after sequentially removing data points based on data valuation from high to low. Lower curves indicate better performance of the data valuation method. Error bars represent the 95% confidence interval of 50 independent experiments.

during progressive data removal. This indicates that OOB-CM can be used for the rapid detection of harmful data points in practical applications.

Next, training data points are progressively removed in descending order based on their data values. After each removal, the remaining data points are used to train a logistic regression model, which is then evaluated on the test dataset. As shown in Fig. 3, the test performance of the model rapidly declines after removing only a small portion of the data points identified by OOB-CM. This result indicates that OOB-CM is effective at detecting critical subsets of data that are highly representative of the target task. By accurately identifying these key data points, OOB-CM enables more efficient data pruning without significantly compromising model performance.

4.3 Learning with Weighted Data Points

Although the ordering of data values can be leveraged to effectively identify different types of data points, reliable data values should also be able to guide model training. Specifically, they should help the model focus more on beneficial data points while reducing the impact of harmful ones. Therefore, we conduct a weighted learning experiment to further evaluate the performance of OOB-CM.

Following the setup in Wang et al. [28], the data values are used as importance weights to train a logistic regression model, which is then evaluated on the test dataset. Therefore, the performance of the logistic regression model reflects the quality of the data values produced by different data valuation methods. As shown in Table 3, although Data Shapley and Beta Shapley demonstrated strong performance in previous experiments, their effectiveness in this experiment is limited. This suggests that these methods are better suited for ranking the importance of training data points rather than directly guiding model training. Consequently, their outputs often require further processing before practi-

Table 3. Accuracy of models trained with weighted samples. The 6 data valuation methods are compared: AME, Beta Shapley (BS), Data-OOB (Data), Data Shapley (DS), LOO, and OOB-CM (ours). 'Uniform' refers to models trained with uniformly weighted samples. Bold numbers indicate the best method. In all cases, the proposed OOB-CM achieves the highest score.

Dataset	Accuracy						
	AME	BS	Data	DS	LOO	ours	Uniform
adult	0.797	0.798	0.791	0.799	0.785	**0.804**	0.782
2dplanes	0.821	0.829	0.825	0.828	0.699	**0.830**	0.821
creditcard	0.783	0.786	**0.790**	0.786	0.739	0.789	0.782
digits	0.791	0.801	0.887	0.796	0.768	**0.891**	0.775
pol	0.850	0.849	0.847	0.849	0.811	**0.856**	0.837
jannis	0.696	0.699	0.695	0.698	0.660	**0.701**	0.683
fried	0.819	0.825	0.826	0.824	0.643	**0.829**	0.825
nomao	0.892	0.893	0.894	0.891	0.875	**0.902**	0.875
CPU	0.904	0.907	**0.912**	0.906	0.749	0.909	0.903
electricity	0.734	0.736	0.732	0.736	0.644	**0.736**	0.729

cal application. In contrast, OOB-CM achieves the best performance, indicating that it not only provides reliable data ranking but can also directly inform and enhance the training process.

4.4 Ablation Analysis

To assess the effectiveness of individual components, we designed an ablation study by progressively removing modules from OOB-CM: (1) OOB-CM, which is the complete version as in previous settings; (2) OOB-CN, which is OOB-CM without multi-round valuation; (3) OOB-NN, which is OOB-CN further without curriculum learning.

As shown in Table 4, a clean training dataset is first created and then cloned into three copies. Noise is added to each copy at different levels: 20%, 30%, and 20%, respectively. The first two noisy datasets are used to train different versions of OOB-CM, while the third serves as the test set. Overall, OOB-CN outperforms OOB-NN, indicating that incorporating curriculum learning in the early stages of weak learner training helps mitigate the negative impact of noisy data. For datasets like 'creditcard', 'nomao', and '2dplanes', which are binary classification tasks. In such tasks, the discriminative information captured by weak learners is sufficiently utilized during the initial round of data valuation. Therefore, multi-round valuation is not necessary. To enable a fairer comparison, we additionally include a multi-class classification dataset, 'covertype'. A comprehensive comparison between OOB-CM and OOB-CN further demonstrates that effectively leveraging the discriminative signals from weak learners benefits model evaluation.

Table 4. F1-score of noisy data detection when the noise level is set to 20% and 30%. The 3 data valuation methods are compared: OOB-CM (ours), OOB-CN (ours-CN), and OOB-NN (ours-NN). Overall, OOB-CM achieves the best performance.

Dataset	F1-score (20%)			F1-score (30%)		
	ours	ours-CN	ours-NN	ours	ours-CN	ours-NN
adult	0.618	0.618	0.611	0.615	0.615	0.583
2dplanes	0.672	0.672	0.666	0.667	0.667	0.613
creditcard	0.603	0.603	0.596	0.595	0.595	0.574
digits	0.846	0.823	0.787	0.801	0.789	0.671
covertype	0.667	0.655	0.616	0.652	0.637	0.602
pol	0.703	0.703	0.688	0.694	0.694	0.655
jannis	0.515	0.515	0.498	0.501	0.501	0.470
fried	0.684	0.684	0.667	0.672	0.672	0.641
nomao	0.811	0.811	0.784	0.797	0.797	0.733
CPU	0.808	0.808	0.779	0.781	0.781	0.739
electricity	0.536	0.536	0.524	0.511	0.511	0.495

5 Conclusion

In this paper, we propose a novel data valuation framework named OOB-CM, which achieves higher utilization of weak learners and improved robustness to noisy data compared to Data-OOB. OOB-CM employs a curriculum learning strategy with dynamic thresholds to select more reliable training data points, thereby protecting weak learners from the adverse effects of noise. Subsequently, it adopts a multi-round voting strategy to iteratively aggregate the prediction signals from weak learners, further enhancing their utility. Experimental results demonstrate that OOB-CM consistently outperforms existing methods on tasks such as noise detection, data removal, and weighted training.

Acknowledgments. This study is partly supported by the Key Research and Development Project of Heilongjiang Province (2022ZX01A34) and the National Natural Science Foundation of China under Grant (60903083).

References

1. Arpit, D., et al.: A closer look at memorization in deep networks. In: International Conference on Machine Learning, pp. 233–242. PMLR (2017)
2. Bengio, Y., Louradour, J., Collobert, R., Weston, J.: Curriculum learning. In: Proceedings of the 26th Annual International Conference on Machine Learning, pp. 41–48 (2009)
3. Cook, R.D., Weisberg, S.: Residuals and Influence in Regression. Chapman and Hall, New York (1982)

4. De Clippel, G., Serrano, R.: Marginal contributions and externalities in the value. Econometrica **76**(6), 1413–1436 (2008)
5. Efron, B.: Jackknife-after-bootstrap standard errors and influence functions. J. Roy. Stat. Soc.: Ser. B (Methodol.) **54**(1), 83–111 (1992)
6. Efron, B., Tibshirani, R.: Improvements on cross-validation: the 632+ bootstrap method. J. Am. Stat. Assoc. **92**(438), 548–560 (1997)
7. Feurer, M., et al.: Openml-python: an extensible python api for openml. J. Mach. Learn. Res. **22**(100), 1–5 (2021). http://jmlr.org/papers/v22/19-920.html
8. Ghorbani, A., Kim, M., Zou, J.: A distributional framework for data valuation. In: International Conference on Machine Learning, pp. 3535–3544. PMLR (2020)
9. Ghorbani, A., Zou, J.: Data shapley: equitable valuation of data for machine learning. In: International Conference on Machine Learning, pp. 2242–2251 (2019)
10. Ieong, S., Shoham, Y.: Marginal contribution nets: a compact representation scheme for coalitional games. In: Proceedings of the 6th ACM Conference on Electronic Commerce, pp. 193–202 (2005)
11. Jia, R., et al.: Efficient task-specific data valuation for nearest neighbor algorithms. Proc. VLDB Endowment **12**(11), 1610–1623 (2019)
12. Jia, R., et al.: Towards efficient data valuation based on the shapley value. In: The 22nd International Conference on Artificial Intelligence and Statistics, pp. 1167–1176. PMLR (2019)
13. Jia, R., et al.: Scalability vs. utility: do we have to sacrifice one for the other in data importance quantification? In: Proceedings of the IEEE/CVF Conference on Computer Vision and Pattern Recognition, pp. 8239–8247 (2021)
14. Jiang, K., Liang, W., Zou, J.Y., Kwon, Y.: Opendataval: a unified benchmark for data valuation. In: Advances in Neural Information Processing Systems, vol. 36 (2023)
15. Just, H.A., et al.: Lava: data valuation without pre-specified learning algorithms. arXiv preprint arXiv:2305.00054 (2023)
16. Kalai, E., Samet, D.: On weighted shapley values. Internat. J. Game Theory **16**(3), 205–222 (1987)
17. Koh, P.W., Liang, P.: Understanding black-box predictions via influence functions. In: International Conference on Machine Learning, pp. 1885–1894. PMLR (2017)
18. Kwon, Y., Zou, J.: Beta shapley: a unified and noise-reduced data valuation framework for machine learning. In: International Conference on Artificial Intelligence and Statistics, pp. 8780–8802. PMLR (2022)
19. Kwon, Y., Zou, J.: Data-oob: out-of-bag estimate as a simple and efficient data value. In: International Conference on Machine Learning, pp. 18135–18152. PMLR (2023)
20. Lin, J., Zhang, A., Lécuyer, M., Li, J., Panda, A., Sen, S.: Measuring the effect of training data on deep learning predictions via randomized experiments. In: International Conference on Machine Learning, pp. 13468–13504. PMLR (2022)
21. Luo, S., Kasaei, H., Schomaker, L.: Accelerating reinforcement learning for reaching using continuous curriculum learning. In: 2020 International Joint Conference on Neural Networks (IJCNN), pp. 1–8. IEEE (2020)
22. Pedregosa, F.: Scikit-learn: machine learning in Python. J. Mach. Learn. Res. **12**, 2825–2830 (2011)
23. Schoch, S., Xu, H., Ji, Y.: Cs-shapley: class-wise shapley values for data valuation in classification. Adv. Neural. Inf. Process. Syst. **35**, 34574–34585 (2022)
24. Shapley, L.S.: A value for n-person games. Contrib. Theory of Games **2**(28), 307–317 (1953)

25. Sun, Y., Shen, J., Kwon, Y.: 2d-oob: attributing data contribution through joint valuation framework. In: The Thirty-Eighth Annual Conference on Neural Information Processing Systems
26. Sundararajan, M., Najmi, A.: The many shapley values for model explanation. In: International Conference on Machine Learning, pp. 9269–9278. PMLR (2020)
27. Tang, S., et al.: Data valuation for medical imaging using shapley value and application to a large-scale chest x-ray dataset. Sci. Rep. **11**(1), 8366 (2021)
28. Wang, J.T., Jia, R.: Data banzhaf: a robust data valuation framework for machine learning. In: Proceedings of the 26th International Conference on Artificial Intelligence and Statistics, vol. 206, pp. 6388–6421 (2023)
29. Wang, X., Chen, Y., Zhu, W.: A survey on curriculum learning. IEEE Trans. Pattern Anal. Mach. Intell. **44**(9), 4555–4576 (2021)
30. Winter, E.: The shapley value. Handb. Game Theory Econ. Appl. **3**, 2025–2054 (2002)
31. Xia, H., Li, X., Pang, J., Liu, J., Ren, K., Xiong, L.: P-shapley: shapley values on probabilistic classifiers. Proc. VLDB Endowment **17**(7), 1737–1750 (2024)
32. Xu, X., Wu, Z., Foo, C.S., Low, B.K.H.: Validation free and replication robust volume-based data valuation. Adv. Neural. Inf. Process. Syst. **34**, 10837–10848 (2021)
33. Yoon, J., Arik, S., Pfister, T.: Data valuation using reinforcement learning. In: International Conference on Machine Learning, pp. 10842–10851. PMLR (2020)
34. Zhou, T., Wang, S., Bilmes, J.: Robust curriculum learning: from clean label detection to noisy label self-correction. In: International Conference on Learning Representations (2021)

Quantum Color Image Selective Encryption in HSI Color Space

Xianhua Song[1](✉), Mengmeng Li[2], and Fuzhou Wang[1]

[1] School of Science, Harbin University of Science and Technology, Harbin 150080, China
songxianhua@hrbust.edu.cn

[2] College of Science, National University of Defense Technology, Changsha 410073, China

Abstract. Regarding security concerns in image storage and transmission, particularly for medical images, we propose a selective encryption method for quantum color images EQIRHSI, which is built on quantum LoG edge detection and quantum morphological processing. Leveraging HSI color space's visual perception characteristics, the algorithm first extracts regions of interest (ROI) via quantum LoG edge detection, then refines ROI boundaries through quantum dilation and erosion. Subsequently, the color information of the ROI is encrypted using a double random phase encryption technique, while the intensity information of the ROI is encrypted by combining quantum walks, 2D-SCLMS chaotic system, XOR, and XNOR operations. Finally, quantum circuits are designed and simulation experiments are implemented, demonstrating that this method can efficiently encrypt and transmit the ROI of the image while ensuring image security. After decryption, the location and information of the ROIs are accurately restored, demonstrating high security and practicality.

Keywords: quantum images · selective encryption · quantum LoG · quantum morphology processing

1 Introduction

With the rapid evolution of the digital era, images have emerged as a crucial medium for information dissemination. Given the numerous security issues involved in transmitting images, selecting appropriate image encryption methods and ensuring that encrypted images remain uncracked during transmission and storage has always been a significant aspect of current information security research [1, 2]. Quantum computing possesses powerful computational capabilities [3–5], and compared to traditional encryption algorithms, the use of quantum encryption algorithms is more effective in protecting images against malicious attacks [6]. Selective Encryption (SE) in image processing [7, 8] is a method capable of performing local perceptual encryption on images. Quantum image selective encryption combines the advantages of quantum computing and image processing, extracting and encrypting the ROI [9, 10] within the image through quantum algorithms. Its purpose is to balance security with computational efficiency, protecting the most visually important parts of the image, thereby reducing unnecessary redundant

X. Zhou et al. (Eds.): GPC 2024, LNCS 15225, pp. 204–227, 2026.
https://doi.org/10.1007/978-981-95-1346-8_13

computations and playing a significant role in fields such as medical imaging and business [11–13]. The above content summarizes the background of image encryption, and points out that the quantum image selective encryption fusion technology provides a solution for the safe and efficient processing of key areas.

Quantum image representation [14–16] serves as the foundational requirement for all quantum image processing operations [6, 17, 18]. Le et al. [19] introduced the Flexible Representation of Quantum Images (FRQI), utilizing normalized superposition to store pixel information and encoding color information with just one qubit. Subsequently, Zhang Yi et al. [20] developed the Novel Enhanced Quantum Representation (NEQR) for digital images, which encodes grayscale values as binary- encoded entangled quantum sequences per pixel. Following the introduction of FRQI and NEQR, a multitude of quantum image representation models have emerged to address various applications. For example, the Multi-Channel Representation for Images on Quantum Computers using the RGBα Color Space (MCRQI) [21] and Multi-Channel Representation for Images (MCQI) [22] are quantum multi-channel representation models extending FRQI to the RGB space.Additional models proposed include the Normal Arbitrary Quantum Superposition State (NAQSS) for multi-dimensional images by Li Haisheng and Zhou Ruigui et al. [23], and Jiang Nan et al.'s [24] Generalized Quantum Image Representation (GQIR), capable of handling images of arbitrary sizes. While most of these models utilize the RGB color space, research suggests the HSI color space better aligns with human color perception [25–28]. Consequently, in 2021, Chen Guanglong et al. [29] devised a quantum color image representation method based on HSI, alongside processing operations leveraging this representation [30, 31]. Subsequently, Li Mengmeng et al. [32] improved the QIRHSI representation method, using two entangled quantum states to represent color information, proposing the EQIRHSI representation model and providing compression and encryption methods for this image.

Selective image encryption techniques in classical images have been proposed by many scholars and applied in fields such as medical imaging. For example, Xiao Di et al. [33] designed a chaotic image encryption algorithm for ROI to encrypt irregular areas. However, it cannot automatically identify ROIs. Ntalianis [34] introduced an automatic human video object encryption scheme utilizing chaos, which uses automatic detection to extract regions of interest (such as faces and bodies) and subsequently encrypts the arranged pixels of these objects with a chaotic system. Building upon this work, Liu Yunan et al. [35] developed a hybrid encryption algorithm which combines chaotic encryption with RSA encryption to enhance security parameters and matching accuracy rates for facial feature template protection. An Gaoyun et al. [36] put forward a face recognition template protection algorithm based on chaotic cryptography technology, and Chen Rong et al. [37] suggested a selective regional medical image encryption algorithm based on cascaded chaos. Prakash et al. proposed a selective encryption method for multidimensional medical images based on chaotic Henon map segmentation masking [38]. Zhang Lina et al. [39] introduced a reversible selective medical image encryption algorithm based on coupled chaotic maps and steganography. However, research on quantum-based selective image encryption algorithms remains in its infancy. Heidari et al. [40] pioneered a novel approach to quantum-selective medical image encryption using BRQI (Bitplane Representation of Quantum Images). This method can effectively encrypt regions of interest in medical images by operating on bitplanes. Subsequently, Hu Hao et al. [41] investigated a quantum region of interest extraction algorithm based on the LC algorithm.

Given the characteristics of the EQIRHSI representation, this paper employs the quantum LoG edge detection method, quantum expansion, and quantum erosion operations to extract the ROI from quantum images. It investigates the use of unitary operators and quantum circuits for selective encryption of quantum color images, specifically targeting the extracted regions of interest, and designs simulation experiments to evaluate the encryption effectiveness. The above methods, combined with quantum image processing technology and selective encryption theory, provide a systematic solution for regional security protection of quantum color images.

2 Related Work

In order to design a reversible selective encryption scheme for medical images that is both secure and efficient, this chapter briefly introduces the knowledge of medical image ROI extraction, encryption and lossless decryption. It mainly includes U^2-Net, Chen chaotic system, Henon chaotic system and run-length encoding. By sorting out these key technologies, the necessary theoretical and technical foundation is laid for the subsequent construction of a complete and reliable reversible selective encryption scheme for medical images.

2.1 EQIRHSI Model

The EQIRHSI is represented by 2n + q + 2 qubits and expresses as follows:

$$\begin{aligned}|I(\theta)\rangle =& \frac{1}{2^{n+\frac{1}{2}}}\sum_{k=0}^{2^{2n}-1}\left(\cos\theta_k^h|00\rangle + \cos\theta_k^s|01\rangle + \sin\theta_k^h|10\rangle + \sin\theta_k^s|11\rangle\right)\\ &\otimes\left|C_k^0C_k^1\cdots C_k^{q-1}\right\rangle\otimes|k\rangle \end{aligned} \quad (1)$$

$|k\rangle$ denotes positional information. $|00\rangle, |01\rangle, |10\rangle, |11\rangle$ represents a four-dimensional computational ground state, θ_k^h, θ_k^s are two angular values encoding hue and saturation information, $C_k^m \in \{0, 1\}$, $m = 0, 1, \cdots, q-1$, $k = 0, 1, ..., 2^{2n-1}$. The positional information (Y-axis and X-axis) needs $2n$ qubits ($y_{n-1}, y_{n-2}, ..., y_0$ and $x_{n-1}, x_{n-2}, ..., x_0$) for encoding. Assuming a quantum image of size $2^n \times 2^n$, the encoding method for positional information is as shown in Eq. (2).

$$\begin{cases}|k\rangle = |y\rangle|x\rangle = |y_{n-1}, y_{n-2}, ..., y_0\rangle|x_{n-1}, x_{n-2}, ..., x_0\rangle\\ y, x \in \left\{0, 1, ..., 2^n - 1\right\}\\ |y_i\rangle|x_i\rangle \in \{|0\rangle, |1\rangle\}, i = 0, 1, ..., n-1\end{cases} \quad (2)$$

The modulus of the quantum image state examined by this method is 1, corresponding to a normalized state. Figure 1 shows an EQIRHSI image of 2 × 2 size, together with its EQIRHSI representation and the corresponding quantum circuit.

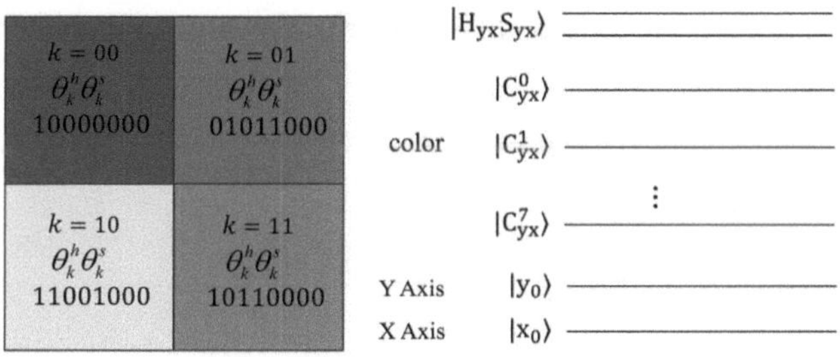

$$
\begin{aligned}
|I\rangle = \frac{1}{2^{3/2}}\{ & (\cos\theta_0^h|00\rangle + \cos\theta_0^s|01\rangle + \sin\theta_0^h|10\rangle + \sin\theta_0^s|11\rangle)|10000000\rangle \otimes |00\rangle \\
& + (\cos\theta_1^h|00\rangle + \cos\theta_1^s|01\rangle + \sin\theta_1^h|10\rangle + \sin\theta_1^s|11\rangle)|01011000\rangle \otimes |01\rangle \\
& + (\cos\theta_2^h|00\rangle + \cos\theta_2^s|01\rangle + \sin\theta_2^h|10\rangle + \sin\theta_2^s|11\rangle)|11001000\rangle \otimes |10\rangle \\
& + (\cos\theta_3^h|00\rangle + \cos\theta_3^s|01\rangle + \sin\theta_3^h|10\rangle + \sin\theta_3^s|11\rangle)|10110000\rangle \otimes |11\rangle\}
\end{aligned}
$$

Fig. 1. The quantum representation and quantum circuit of EQIRHSI

2.2 Quantum LoG

The Laplacian of Gaussian (LoG) operator is a second-order differential operator. For suppressing noise, a Gaussian filter is selected to filter and denoise the quantum image. Then, second-order directional derivative of the image pixel values is calculated, and the zero-crossing points or extreme value positions of the second-order directional derivative are searched to obtain the edges of the quantum image. The Gaussian filter mask with a size of $(2k + 1) \times (2k + 1)$ can be expressed as:

$$
w = \begin{bmatrix}
h_{Y-k,X-k} & h_{Y-k,X-k+1} & \cdots & h_{Y-k,X+k} \\
h_{Y-k+1,X-k} & h_{Y-k+1,X-k+1} & \cdots & h_{Y-k+1,X+k} \\
\vdots & \vdots & & \vdots \\
h_{Y,X-k} & h_{Y,X-k+1} & \cdots & h_{Y,X+k} \\
\vdots & \vdots & & \vdots \\
h_{Y+k,X-k} & h_{Y+k,X-k+1} & \cdots & h_{Y+k,X+k}
\end{bmatrix} \tag{3}
$$

The expression in Eq. (3) can be calculated using the expression in Eq. (4):

$$
G''(x, y) = \frac{1}{\sigma^2}(\frac{x^2 + y^2}{\sigma^2} - 1)e^{-(x^2+y^2)/2\sigma^2} \tag{4}
$$

2.3 Quantum Morphological Operations

As fundamental operations in morphological image processing, dilation and erosion form its foundational basis. By selecting specific structural elements and performing operations with the extracted image, the goals of image recognition and analysis can be achieved. Suppose a binary image A of size $2^n \times 2^n$, and a structural element B of size

$(2k + 1) \times (2k + 1)$, with its origin at the center of the structural element, the dilation operation is:

$$A \oplus B = \{z|(B)_z \cap A \neq \varnothing\} \tag{5}$$

The erosion operation is:

$$A \Theta B = \left\{z|(B)_z \cap A^c = \varnothing\right\} \tag{6}$$

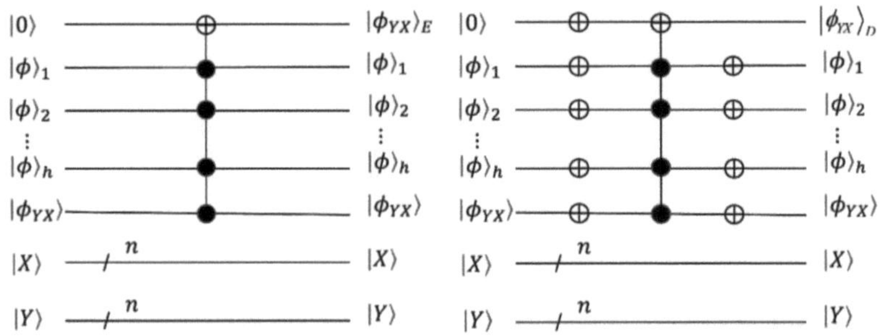

Fig. 2. Circuit diagram of quantum corrosion and quantum expansion

The definition of quantum binary morphological erosion for the EQIRHSI representation model is as follows:

$$|\phi_{YX}\rangle_E = \frac{1}{2^{n+1}} \sum_{Y=0}^{2^n-1} \sum_{X=0}^{2^n-1} \left(|\phi_{YX}\rangle \cap |\phi\rangle_1 \cap |\phi\rangle_2 \cap \cdots \cap |\phi\rangle_h\right) \otimes |Y\rangle|X\rangle \tag{7}$$

In this context, represents the binary information of each point in the binary image and the corresponding neighborhood pixels in the structuring element, excluding the origin. Quantum binary morphological dilation can be defined as:

$$|\phi_{YX}\rangle_D = \frac{1}{2^{n+1}} \sum_{Y=0}^{2^n-1} \sum_{X=0}^{2^n-1} \left(|\phi_{YX}\rangle \cup |\phi\rangle_1 \cup |\phi\rangle_2 \cup \cdots \cup |\phi\rangle_h\right) \otimes |Y\rangle|X\rangle \tag{8}$$

The quantum circuit diagram for binary morphological erosion and dilation is shown in Fig. 2, where $|\phi_{YX}\rangle_E$ and $|\phi_{YX}\rangle_D$ represent the pixel information for quantum erosion and quantum dilation results, respectively.

3 Selective Encryption and Decryption of EQIRHSI Image

The selective encryption of EQIRHSI images involves two steps. Firstly, the intensity channel of the EQIRHSI image undergoes quantum LoG operation, quantum erosion, and quantum dilation to obtain the coordinate information of the ROI. Subsequently, the position of the ROI in the original image is encrypted, ultimately resulting in a selectively encrypted image. Figure 3 illustrates the flowchart of the selective encryption algorithm for quantum color images based on EQIRHSI. The above steps and diagrams clearly show the complete process of the selective encryption.

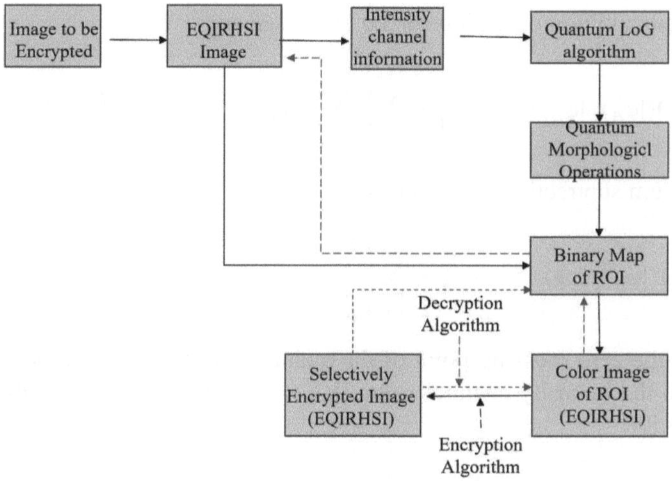

Fig. 3. Flow diagram of selective encryption algorithm for quantum color images

3.1 EQIRHSI Region of Interest Extraction

The representation of EQIRHSI images is shown in Eq. (1), and the steps for extracting the region of interest are:

Step 1: Preparing h identical quantum states $|\varphi_{YX}\rangle$, which can be formulated as

$$\left\{|\varphi_{YX}\rangle_1, |\varphi_{YX}\rangle_2, \cdots, |\varphi_{YX}\rangle_h\right\} \tag{9}$$

h is obtained by the following formula:

$$h = \left|h_{Y-k,X-k}\right| + \left|h_{Y-k,X-k+1}\right| + \cdots + \left|h_{Y,X}\right| + \cdots + \left|h_{Y+k,X+k}\right| \tag{10}$$

Step 2: Perform a shift operation on h quantum image states to obtain a series of new quantum image states, storing the original image and its neighboring information.

$$\{h_{Y-k,X-k}|\varphi_{Y-k,X-k}\rangle, h_{Y-k,X-k+1}|\varphi_{Y-k,X-k+1}\rangle, \cdots, h_{Y,X}|\varphi_{YX}\rangle, \cdots h_{Y+k,X+k}|\varphi_{Y+k,X+k}\rangle\} \tag{11}$$

$h_{i,j}$ is the weight coefficient, it is necessary to prepare $h_{i,j}\ |\varphi_{YX}\rangle$ to facilitate quantum circuit operations.

Step 3: Quantum image addition and subtraction operations for mask-image convolution are implemented as follows. The product of the structural elements and the neighboring pixels is transformed into the multiple addition process of the neighboring pixels themselves, and the quantum circuit diagram of the quantum shift operation is introduced in Fig. 4. Given two quantum states:

$$\begin{aligned} |\varphi_A\rangle &= \frac{1}{2^{n+1/2}} \sum_{Y=0}^{2^n-1} \sum_{X=0}^{2^n-1} |H_{YX} S_{YX}\rangle \left|I_{YX}^A\right\rangle \otimes |Y\rangle|X\rangle \\ |\varphi_B\rangle &= \frac{1}{2^{n+1/2}} \sum_{Y=0}^{2^n-1} \sum_{X=0}^{2^n-1} |H_{YX} S_{YX}\rangle \left|I_{YX}^B\right\rangle \otimes |Y\rangle|X\rangle \end{aligned} \tag{12}$$

Its quantum addition operation can be expressed as:

$$\mathrm{QADD}(|\varphi_A\rangle, |\varphi_B\rangle) = \frac{1}{2^{n+1/2}} \sum_{Y=0}^{2^n-1} \sum_{X=0}^{2^n-1} |H_{YX} S_{YX}\rangle \left| I_{YX}^A + I_{YX}^B \right\rangle \otimes |Y\rangle |X\rangle \tag{13}$$

The quantum subtraction operation is

$$\mathrm{QSUB}(|\varphi_A\rangle, |\varphi_B\rangle) = \frac{1}{2^{n+1/2}} \sum_{Y=0}^{2^n-1} \sum_{X=0}^{2^n-1} |H_{YX} S_{YX}\rangle \left| I_{YX}^A - I_{YX}^B \right\rangle \otimes |Y\rangle |X\rangle \tag{14}$$

Step 4: Find the zero crossing point of the quantum state, that is the position $|k\rangle$ corresponding to the intensity channel value $|I_k\rangle' = 0$, set the auxiliary qubit as $|d\rangle$ and initialize it as $|0\rangle$:

$$\begin{aligned} |\varphi\rangle' &= |0\rangle \otimes |\varphi\rangle \\ &= |0\rangle \otimes \frac{1}{2^{n+1/2}} \sum_{k=0}^{2^{2n}-1} |H_k S_k\rangle |I_k\rangle' \otimes |k\rangle \\ &= \frac{1}{2^{n+1/2}} \sum_{k=0}^{2^{2n}-1} |H_k S_k\rangle |0\rangle |I_k\rangle' \otimes |k\rangle \\ &= \frac{1}{2^{n+1/2}} \sum_{k=0}^{2^{2n}-1} |H_k S_k\rangle |I_k\rangle \otimes |p\rangle \otimes |k\rangle \end{aligned} \tag{15}$$

$|I_k\rangle' = 0$, $|k\rangle$ is the edge of the image, at this time the auxiliary qubit $|d\rangle$ output $|p\rangle = |1\rangle$, otherwise, $|p\rangle = |0\rangle$, it represents the edge information. In Fig. 5, a quantum circuit diagram is designed to find edge information.

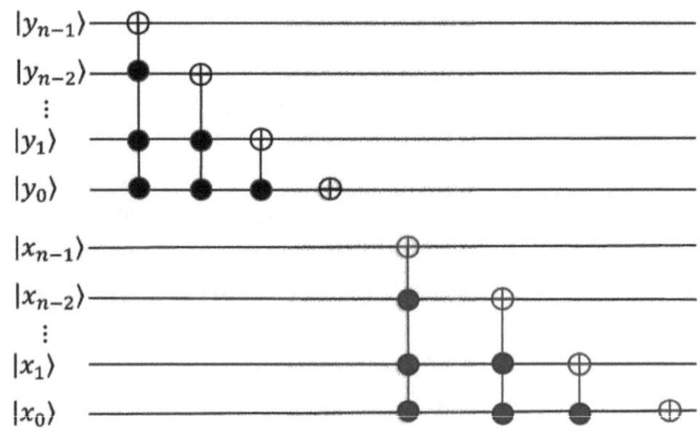

Fig. 4. Diagram of a quantum cyclic shift circuit in the *x*, *y* direction

Step 5: Note that the quantum image after **Step 4** operation is:

$$|\phi\rangle = \frac{1}{2^{n+1/2}} \sum_{k=0}^{2^{2n}-1} |H_k S_k\rangle |I_k\rangle \otimes |p\rangle \otimes |k\rangle \tag{16}$$

In order to extracts the ROI information better, the image after edge extraction is processed by quantum morphology, and its structural elements are set as:

$$\begin{pmatrix} 0 & 1 & 0 \\ 1 & 1 & 1 \\ 0 & 1 & 0 \end{pmatrix} \tag{17}$$

Preparing $h = 4$ identical $|\phi\rangle$ and performing quantum shift operation to obtain the product.

$$\left\{ \left|\phi_{Y,X-1}\right\rangle, \left|\phi_{Y,X+1}\right\rangle, \left|\phi_{Y,X}\right\rangle, \left|\phi_{Y-1,X}\right\rangle, \left|\phi_{Y+1,X}\right\rangle \right\} \tag{18}$$

For convenience, corrosion followed by expansion is regarded as one operation. Add:

$$\{|\phi_{Y-1,X-1}\rangle, |\phi_{Y+1,X-1}\rangle, |\phi_{Y-1,X+1}\rangle, |\phi_{Y+1,X+1}\rangle, |\phi_{Y-2,X}\rangle, |\phi_{Y+2,X}\rangle, |\phi_{Y,X-2}\rangle, |\phi_{Y,X+2}\rangle\} \tag{19}$$

The quantum circuit is shown in Fig. 6. $\left|C_P^{i,j}\right\rangle$ represents the value at position $|i\rangle|j\rangle$ of $|p\rangle$ after $\left|\phi_{i,j}\right\rangle$ is shifted, and $\left|C_P^{X,Y}\right\rangle_0$ denotes the binary information of the image after edge extraction. The regions with a pixel value of $|1\rangle$ are the regions of interest.

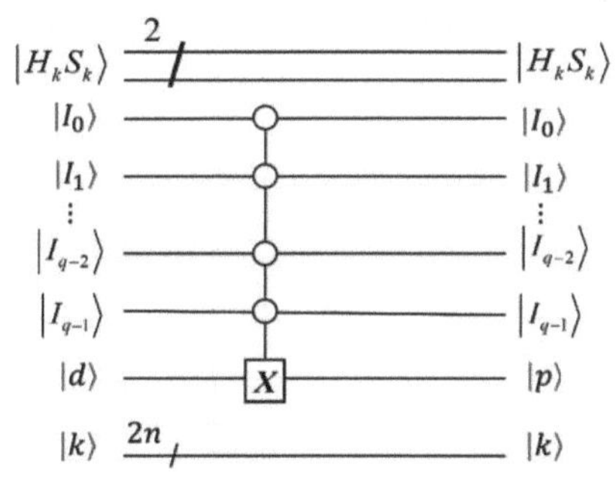

Fig. 5. Circuit diagram of quantum LoG edge detection

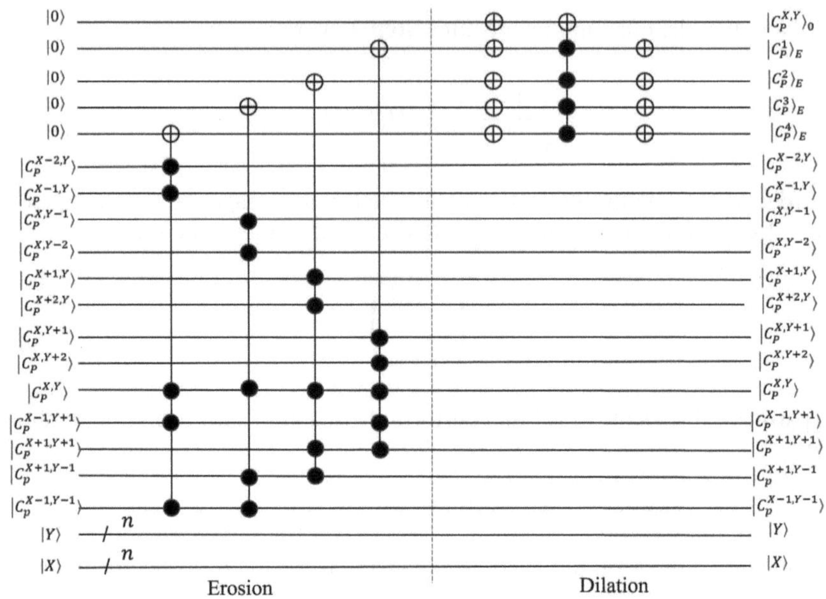

Fig. 6. Diagram of the quantum morphology operation circuit

3.2 EQIRHSI Selective Encryption and Decryption

The extracted region of interest is encrypted using a method that performs double random phase encryption on the entangled hue and saturation channels, and uses bit plane cross-swapping, XOR, and equivalence operations for diffusion on the intensity channel. The encryption key sequence is generated by quantum random walk and 2D-SCLMS chaotic sequence, with the specific encryption method detailed in Reference [42].

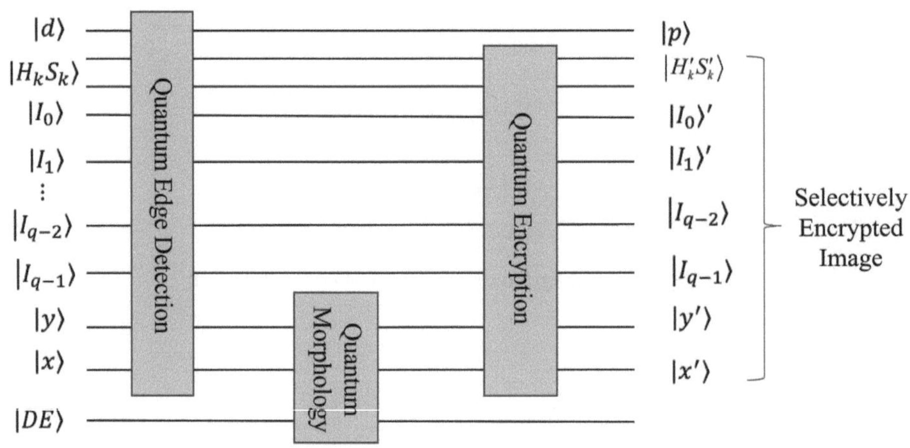

Fig. 7. Circuit diagram of Selective encryption

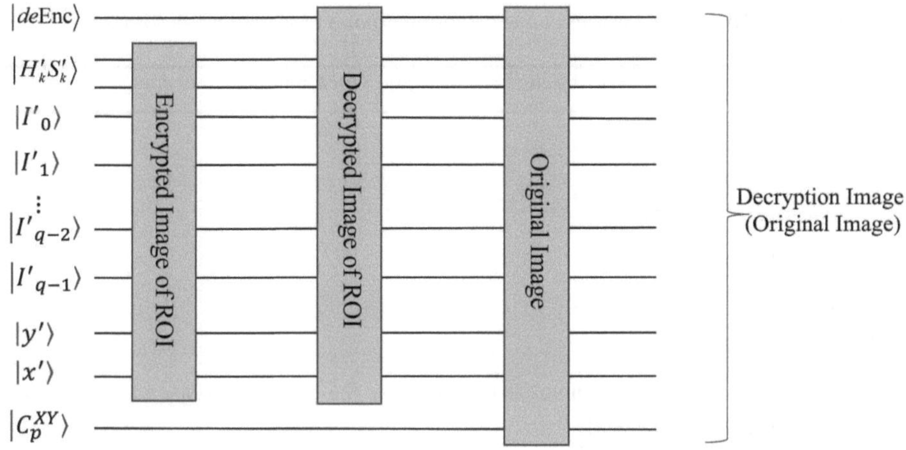

Fig. 8. Circuit diagram of decryption

Figure 7 depicts the quantum circuit diagram for selective encryption, where Quantum Circuit $|DE\rangle$ represents the quantum morphological operations performed on the image after quantum edge detection. The specific process of the selective encryption algorithm for quantum color images is shown in Table 1.

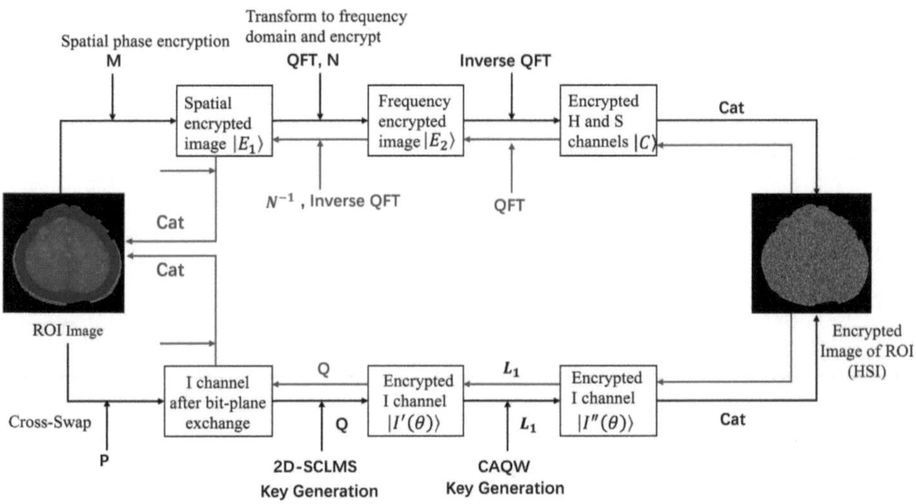

Fig. 9. Encrypt and decrypt the extracted images of the ROI [42]

Input Information: Quantum color image $|\varphi_{YX}\rangle$ and auxiliary qubits $|d\rangle$,$|0\rangle$;

Step 1: Perform a convolution operation on the intensity channel (represented by a quantum bit sequence) with a Gaussian filter mask (using quantum addition and subtraction

Table 1. The specific process of selective encryption of quantum color images

Algorithm: The Specific Process of Selective Encryption for Quantum Color Images

Input:

1. $|\varphi_{YX}\rangle = \frac{1}{2^{n+1/2}} \sum_{Y=0}^{2^n-1} \sum_{X=0}^{2^n-1} |H_{YX} S_{YX}\rangle |I_{YX}\rangle \otimes |Y\rangle |X\rangle$
2. $|d\rangle = |0\rangle$
3. $|0\rangle$

Process:

Step 1: Prepare h identical quantum states $|\varphi_{YX}\rangle$ for the convolution operation between the intensity channel and the mask.

Step 2: Locate zero-crossing points of quantum states to obtain the quantum color image $|\phi_{YX}\rangle$ containing edge information $|p\rangle$.

Step 3: Perform quantum morphological processing on $|\phi_{YX}\rangle$ to obtain the binary map $|C_P^{X,Y}\rangle_0$.

Step 4: Traverse the coordinate information (i, j) of the image.

if $|C_P^{i,j}\rangle_0 = 1$;

$|\varphi_{i,j}\rangle = \text{Enc}|\varphi_{i,j}\rangle$;

Step 5: Obtain the encrypted image $|\varphi_{YX}\rangle_{\text{Enc}}$ of the region of interest.

Output:

1. $|p\rangle$
2. $|C_P^{X,Y}\rangle_0$
3. $|\varphi_{YX}\rangle_{\text{Enc}}$

operations) to obtain preliminary binary information $|p\rangle$ of the region of interest (see Fig. 10(d));

Step 2: Apply quantum morphological operations to the image $|\phi_{YX}\rangle$ after quantum LoG edge detection, including quantum erosion and dilation, to generate a binary map $|C_P^{X,Y}\rangle_0$ of the region of interest (see Fig. 10(e));

Step 3: Extract binary information where the white regions in Fig. 10(f) represent the region of interest;

Step 4: Traverse the position information (i, j) of the image. If $\left|C_P^{i,j}\right\rangle_0 = 1$, then $|\varphi_{i,j}\rangle$ is identified as part of the region of interest in the quantum color image;
Step 5: Apply the encryption algorithm from Fig. 9 [42] to the region of interest, encrypting the hue, saturation, and intensity channels separately to obtain the encrypted image in HSI color space (see Fig. 10(h)). Convert this to RGB color space, resulting in the selectively encrypted quantum color image shown in Fig. 10(i);

Output Information: Quantum LoG edge detection results $|p\rangle$, binary ROI map $\left|C_P^{X,Y}\right\rangle_0$, and selectively encrypted quantum color image $|\varphi_{YX}\rangle_{\text{Enc}}$.

Figure 8 illustrates the corresponding decryption method. Here, $|de\text{Enc}\rangle$ represents the quantum circuit for the decryption method, and $\left|C_p^{XY}\right\rangle$ denotes the binary map information of the ROI. First, the inverse process of encryption is used to obtain the decrypted ROI information, i.e., the selective decryption graph. Then, the decrypted information is embedded into the selective encryption graph based on the ROI binary map information to recover the original image.

4 Analysis of Simulation Results

A good encryption algorithm should resist various attacks. In this paper, a medical image of the Brain (255 × 255) was selected as the processing target to evaluate the security of the encryption algorithm. The following analyses were conducted: statistical and differential analysis, key space analysis, key sensitivity analysis, anti-cropping attack analysis, and complexity analysis. The above multi-dimensional security analysis comprehensively verifies the stability and reliability of the encryption algorithm against various attacks, and provides a reference for its practical application in the field of image security.

4.1 Statistical and Differential Analysis

(1) **Histogram Analysis**

Experiments on the histograms, information entropy, correlation, and spectrograms of the original and encrypted images were conducted to analyze the encryption algorithm's capability to resist statistical attacks. The fewer statistical information an image reveals, the safer the encryption scheme is. As shown in Fig. 11 A)–D), they represent the original Brain image and the histograms of each channel. Figure 11 a)–d) show the corresponding histograms after encryption. It can be seen from the figures that the histograms of each channel after encryption exhibit more uniformly distributed compared to those of the initial channels. This indicates that the proposed encryption method possesses superior performance in resisting statistical attacks.

(2) **Correlation Analysis**

To analyze the effectiveness of encryption, we utilized the correlation coefficient between adjacent pixels. The correlation coefficient ranges from $[-1, 1]$, where values less than 0 indicate negative correlation and values greater than 0 signify

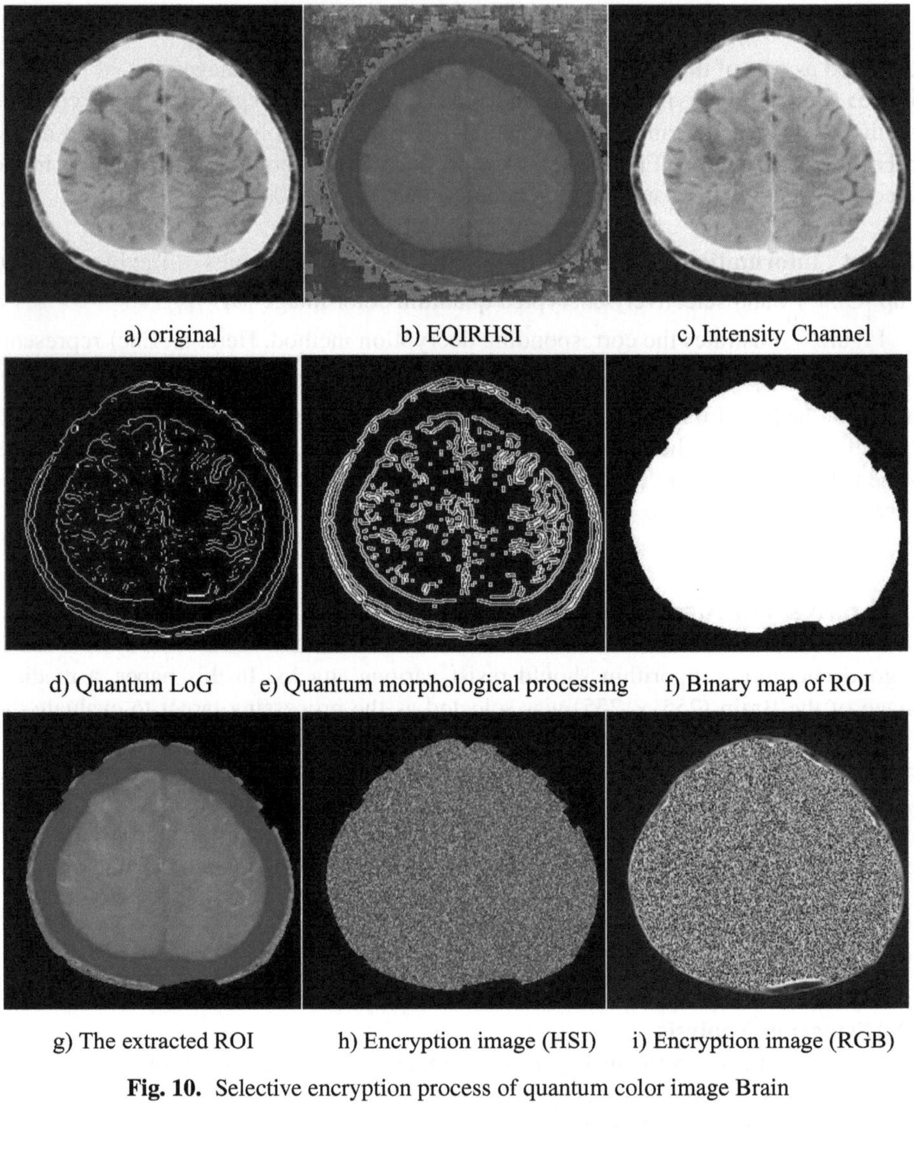

Fig. 10. Selective encryption process of quantum color image Brain

positive correlation. The correlation coefficient is closer to 0 implies the smaller the correlation and the stronger the anti-attack capability of the encryption algorithm. A total of $N = 10000$ pairs of pixels were randomly selected from the Lena image, and the correlation of adjacent pixels (x, y) was calculated horizontally, vertically,

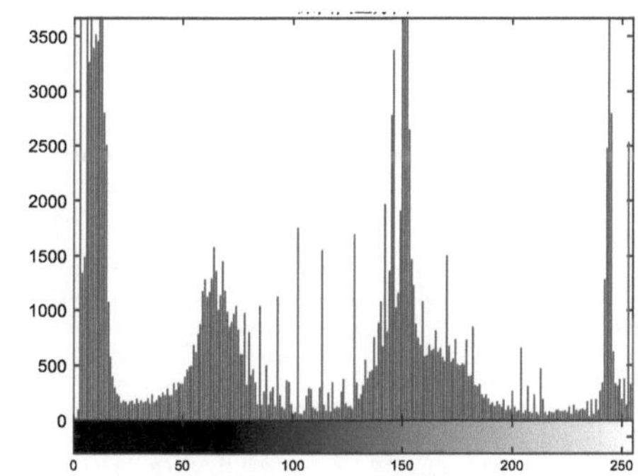

A) Original image

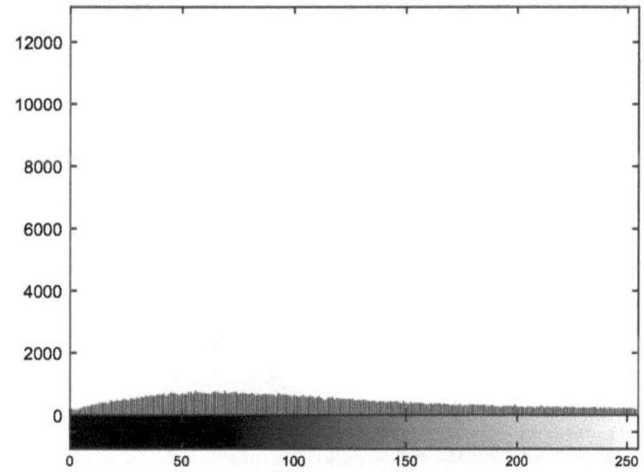

a) Encrypted image

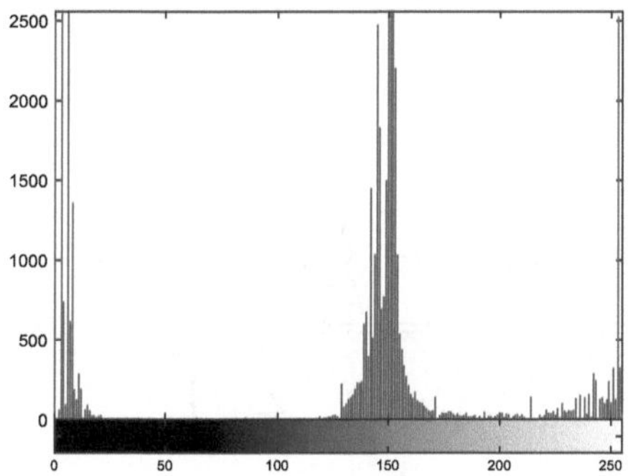

B) The H-channel of original image

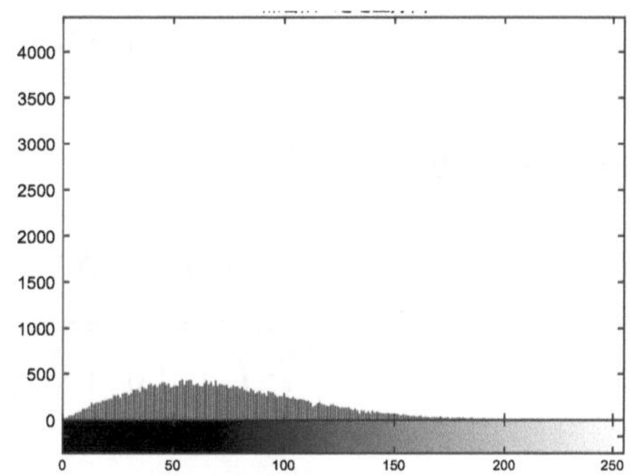

b) The H-channel of the encrypted image

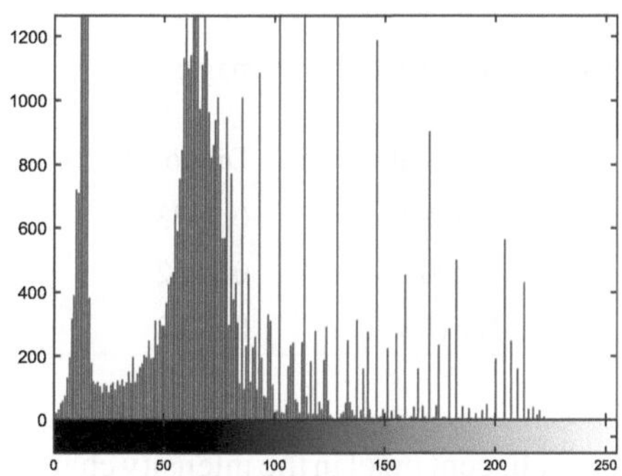

C) The S-channel of original image

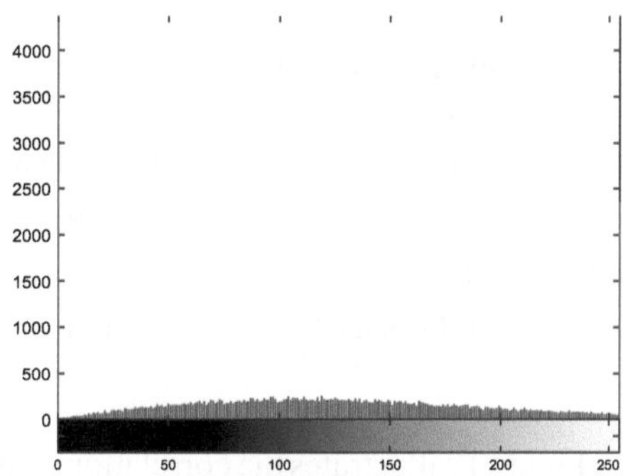

c) The S-channel of the encrypted image

Fig. 11. Histogram of the Brain before and after encryption

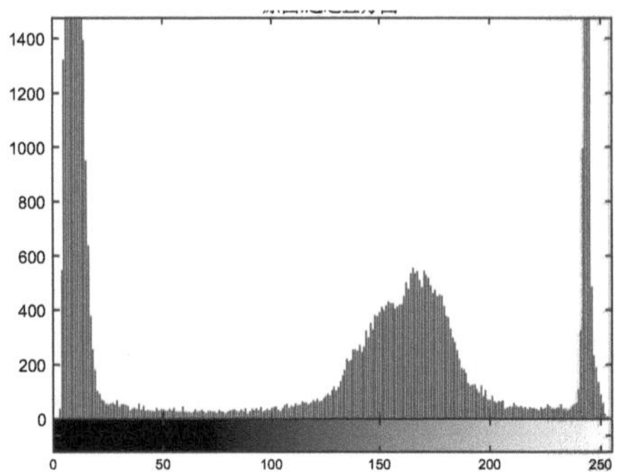

D) The I-channel of original image

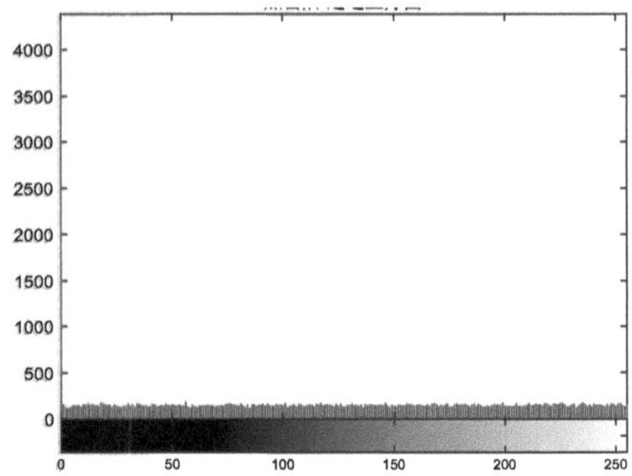

d)The I-channel of the encrypted image

Fig. 11. *(continued)*

and diagonally using Eq. (20).

$$\gamma_{xy} = \frac{\sum_{i=1}^{N}\left(x_i - N^{-1}\sum_{i=1}^{N}x_i\right)\left(y_i - N^{-1}\sum_{i=1}^{N}y_i\right)}{\sqrt{\sum_{i=1}^{N}\left(x_i - N^{-1}\sum_{i=1}^{N}x_i\right)^2}\cdot\sqrt{\sum_{i=1}^{N}\left(y_i - N^{-1}\sum_{i=1}^{N}y_i\right)^2}} \tag{20}$$

Here, γ_{xy} denotes the correlation coefficient between adjacent pixels x and y.

Table 2. Correlation coefficients for the original and encrypted images

Image	Horizontal	Vertical	Diagonal
Image Brain	0.9750	0.9780	0.9673
Brain (I Channel)	0.9897	0.9900	0.9852
Encrypted Brain (I Channel)	0.6138	0.6154	0.6182

Figure 12 illustrates the correlations among adjacent pixel in the intensity channel of the Brain image before and after encryption. Table 2 provides the specific values of these correlation coefficient for both the original image and the ciphertext image in their intensity channel. Notably, after encryption, this intensity channel's correlation is closer to 0 compared to before encryption. Since the analysis is performed on the intensity channel of the entire image except for the region of interest (ROI), the correlation value of adjacent coefficients is relatively large. Nevertheless, it remains evident that the encryption method exhibits commendable performance in against correlation attacks.

(3) **NPCR Analysis**

The Number of Pixels Change Rate (NPCR) serves as an important metric for assessing the difference between two different encrypted images. When a differential

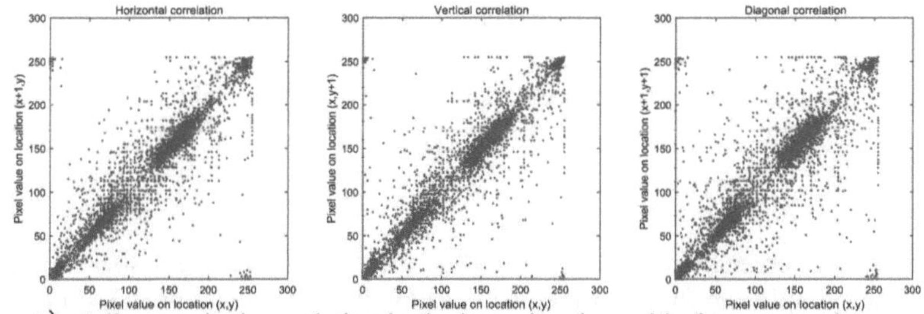

A) Adjacent pixel correlation in the intensity channel before encryption

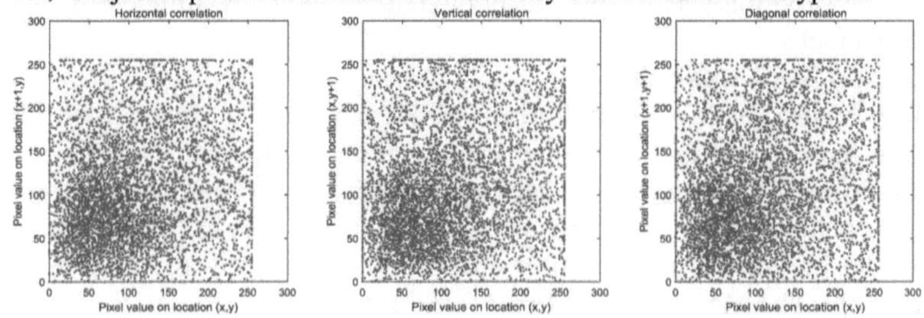

a) Adjacent pixel correlation in the intensity channel after encryption

Fig. 12. The correlation coefficients between primary and encrypted images on intensity channel

attack is performed on an image, a higher NPCR value indicates that it is more difficult for an attacker to crack the ciphertext by altering the plaintext image, and the encryption scheme has stronger resistance to differential attacks.

$$\text{NPCR} = \frac{\sum_{i,j} D(i,j)}{M \times N} \times 100\% \tag{21}$$

Here, M and N represent the length and width of the image to be encrypted, respectively, and $D(i, j)$ is defined as follows:

$$D(i,j) = \begin{cases} 1, & C_1(i,j) \neq C_2(i,j) \\ 0, & C_1(i,j) = C_2(i,j) \end{cases} \tag{22}$$

$C_1(i, j)$ represents the ciphertext image obtained by encrypting the original plaintext image, and $C_2(i, j)$ denotes the ciphertext image obtained by altering a single pixel of the original plaintext image.

Through analysis of Table 3, since the NPCR analysis was performed on the intensity channel of the entire image except for the region of interest (ROI), although the NPCR values of each channel are significantly lower than the ideal value of 99.60%, they all approach 99.00%, particularly for the I channel. Therefore, the encryption method demonstrates relatively superior performance in resisting differential attacks.

(4) **Spectral Analysis**

Spectral analysis is frequently employed as a criterion to evaluate the statistical performance of ciphertext images. Figure 13 shows the spectrograms of Brain

Table 3. NPCR values of encrypted Brain

Image	H Channel	S Channel	I Channel
Brain	87.83%	85.80%	89.21%

before and after encryption. Compared to the original image's spectrogram with significant amplitude fluctuations, the encrypted image displays markedly uniform amplitude variations. This spectral homogeneity indicates effective resistance to spectral attacks.

(5) **NPCR and MSE Analysis**

The Peak Signal-to-Noise Ratio (PSNR) measures the pixel-wise difference between the original and encrypted images, typically used to evaluate encryption quality. A larger PSNR value indicates a greater discrepancy between the two images, implying better encryption effectiveness. For an image to be encrypted with size $2^8 \times 2^8$, let the images before and after encryption be denoted as $X(i,j)$ and $Y(i,j)$, respectively. The Mean Squared Error (MSE) is:

$$\text{MSE} = \frac{1}{2^n \times 2^n} \sum_{i=0}^{2^n-1} \sum_{j=0}^{2^n-1} \{X(i,j) - Y(i,j)\}^2 \tag{23}$$

The PSNR calculation formula is:

$$\text{PSNR} = 10\log_{10}^{\left(\frac{2^8-1}{\sqrt{\text{MSE}}}\right)^2} \tag{24}$$

Table 4 presents both the MSE and PSNR values across each channel of the Brain image, indicating that this encryption method showcases superior encryption performance in term of security.

4.2 Key Space Analysis

Analyzing the key space associated with an encryption scheme can prevent attackers from using brute-force decryption techniques. According to current computational capabilities, an ideal key space should exceed 2^{100} [43]. When decrypting the H and S channels, the encryption keys $\varphi_k^h, \varphi_k^s, \phi_k^h$ and ϕ_k^s are both real numbers within the range [0, 2π]. By applying quantum walks on a circle, a pseudorandom sequence with parameters $(message, N, c, \theta_0, \theta_1, \theta_2)$ are generated, and the initial parameters of the 2D-SCLMS chaotic system for key generation are $x_0', y_0', u', z_0', w_0', \alpha'$ [44]. In the experimental test, a binary string "message" of length 40 is used; thus, the key space of "message" is 2^{40}. Considering the computational precision of other control parameters is 10^{-15}, the key space of this encryption scheme as follows:

$$\text{Key Space} = 2^{40} \times \left(10^{15}\right)^{11} \approx 10^{177} \approx 2^{588} \gg 2^{100} \tag{25}$$

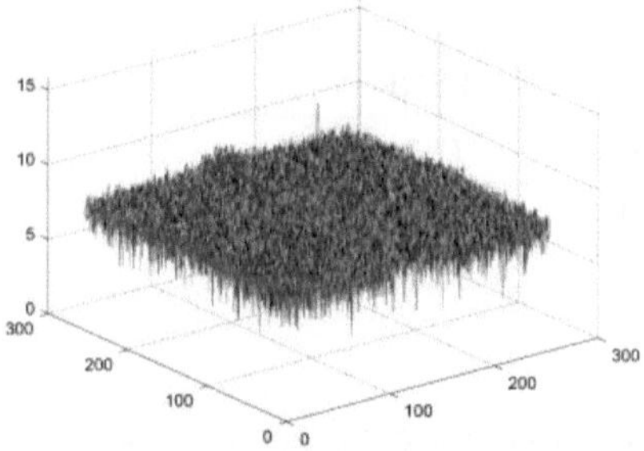

A) Brain (H)

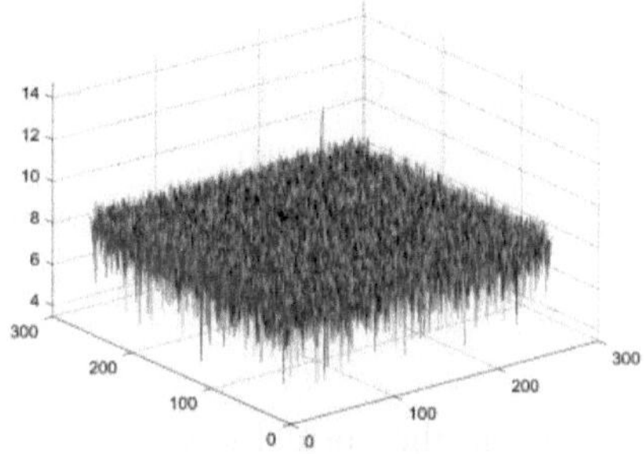

a) Enc Brain (H)

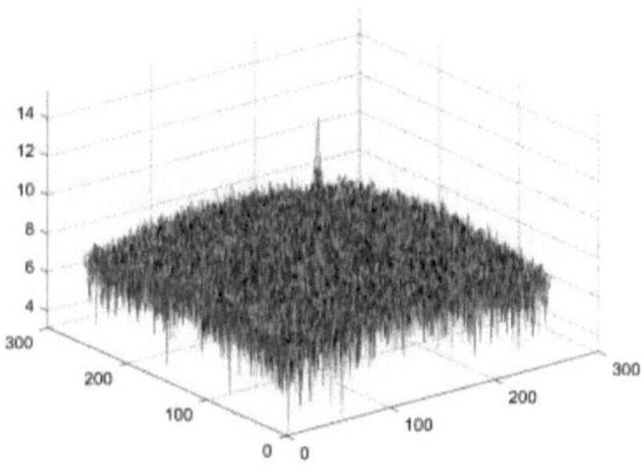

B) Brain (S)

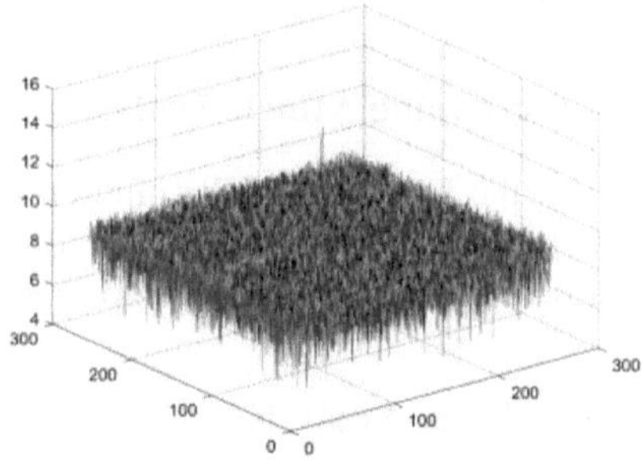

b) Enc Brain (S)

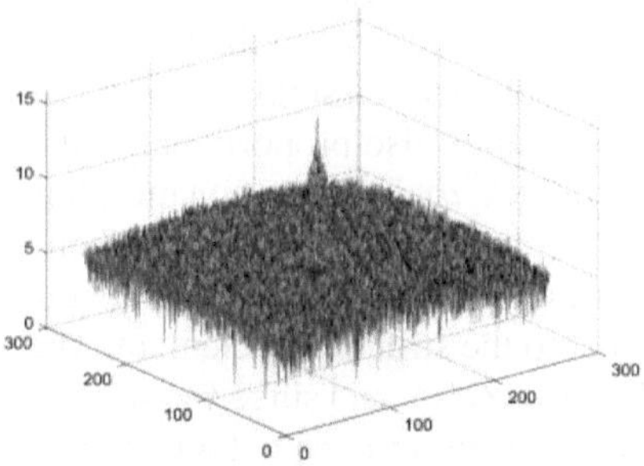

C) Brain (I)

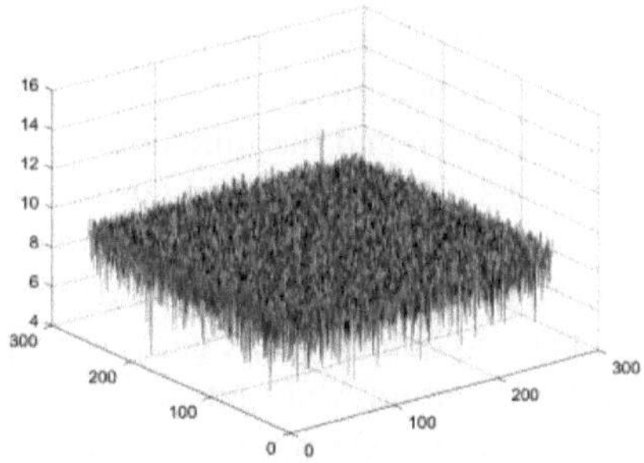

c) Enc Brain (I)

Fig. 13. The spectrogram of Brain before and after encryption

4.3 Key Sensitivity Analysis

Key sensitivity analysis plays an important role in evaluating encryption algorithms. Slight key modifications preventing successful decryption confirm the encryption scheme's efficacy. According to Fig. 14: a), b), c) show decrypted images obtained by altering the key in the first encryption stage. d), e), f) represent decrypted images obtained by changing the key in the second encryption stage. h), i), j) display decrypted images

Table 4. MSE and PSNR values for each channel of the Brain

Channel	MSE	PSNR
Brain (H)	1.1466×10^4	7.5368
Brain (S)	1.3659×10^4	6.7766
Brain (I)	7.0022×10^3	9.6785

obtained using the correct key. Only through utilizing the correct key can one fully restore the original image. This characteristic illustrates that our proposed algorithm demonstrates high sensitivity to alterations in keys while exhibiting robust resistance against key-based attacks.

4.4 Anti-cropping Attack Analysis

In Fig. 15, the I channel of the ciphertext image Brain is occluded by 25%, 50%, and 75%, respectively, and the unoccluded parts can still be decrypted with the key. This encryption method demonstrates good resistance to cropping attacks.

4.5 Complexity Analysis

This section discusses the time complexity required for selective encryption of quantum color images, taking an image of size $2^n \times 2^n$ as an example. The time complexity of executing a single quantum shift operation is $O(n^2)$. A total of $(2k + 1) \times (2k + 1)$ shift operations are required, which is proportional to the mask size, so the total time complexity for shift operations is $O(k^2n^2)$. The time complexity of a single quantum addition operation is $O(q)$, and the time complexity of this step is also proportional to the mask size $(2k+1) \times (2k+1)$, resulting in a complexity of $O(k^2q)$. For quantum morphological processing, the time complexity of shift operations is approximately $O(n^2)$. When implementing quantum binary dilation and erosion operations, the required number of quantum AND gates and quantum OR gates is proportional to the structuring element. The complexity of a quantum AND gate or quantum OR gate is $O(1)$, and since $k = 4$ in quantum morphological processing, the time complexity for quantum morphological processing is $O(1)$. In summary, the time complexity required for selective encryption of quantum color images is approximately: $O(k^2n^2)+O(k^2q)+O(n^2)+O(1) = O(k^2n^2)+O(k^2q)$.

5 Conclusion

This paper proposes a selective encryption method for color images using quantum LoG and quantum morphology. Built upon the EQIRHSI representation method, the approach employs a quantum LoG edge detection operator, quantum erosion, and quantum dilation operations to design corresponding quantum circuits for extracting ROI in quantum images. The extracted ROI is then encrypted using a specially designed encryption algorithm to generate the final selectively encrypted image. Experimental simulations demonstrate the feasibility of the algorithm. Compared with other methods,

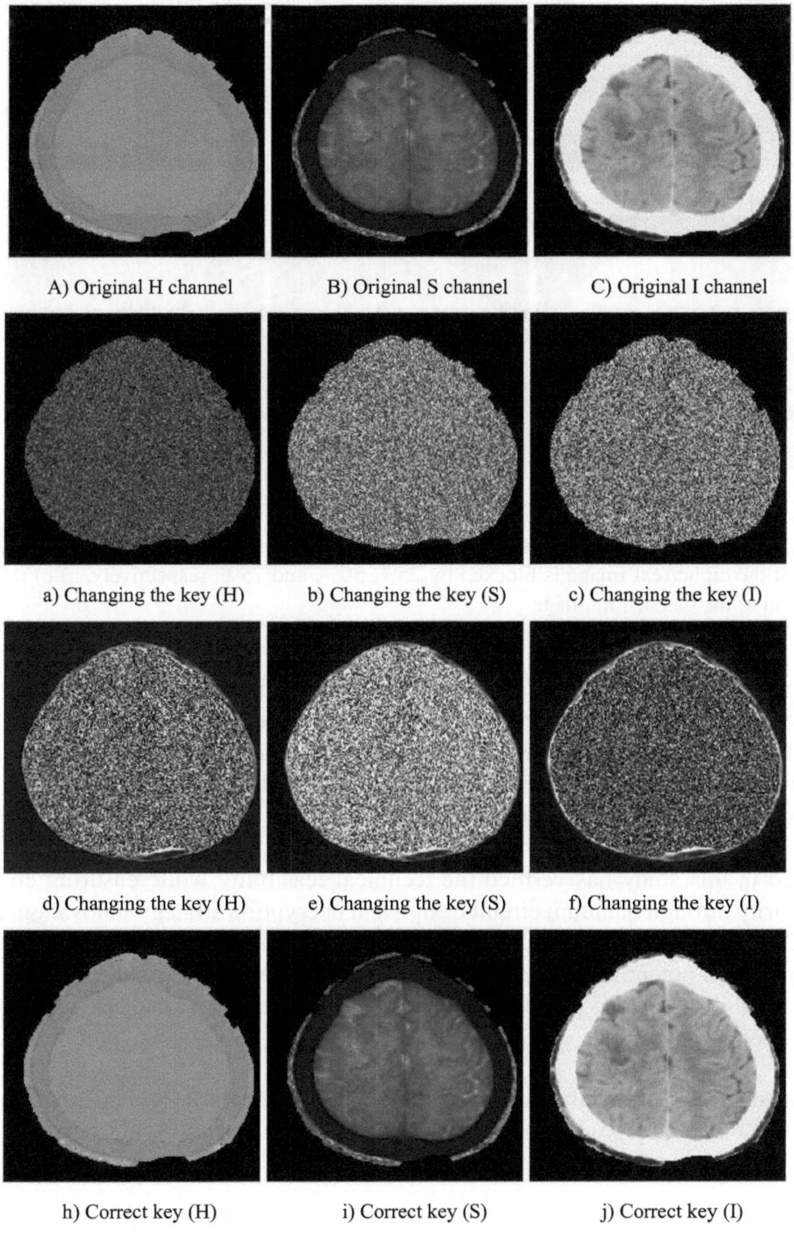

Fig. 14. Changing the key and conducting sensitivity analysis. A), B), and C) respectively represent the H, S, and I channels of plaintext images, a), b), c), d), e), and f) show images decrypted by changing the key of phase encryption, and h), i) and j) are images decrypted with the correct key

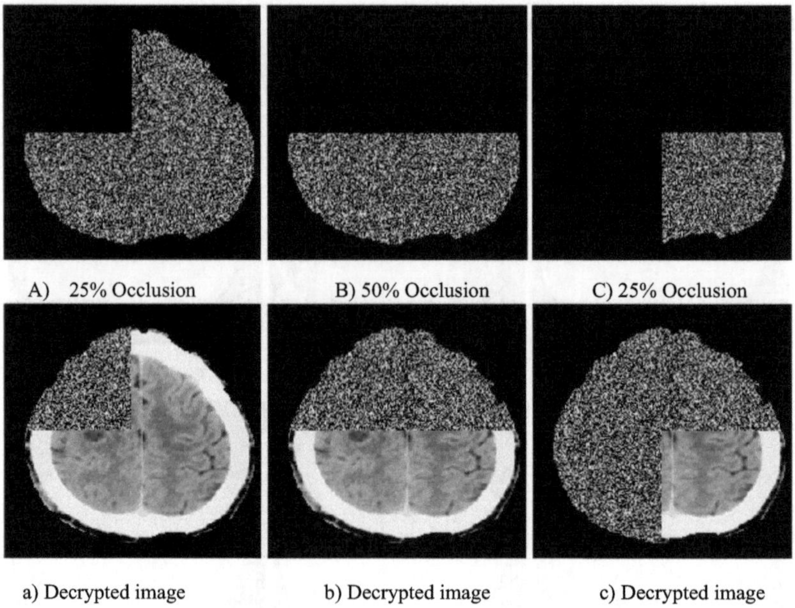

Fig. 15. The ciphertext image tree's intensity channel was analyzed for anti-cropping. A)-C) show that the ciphertext image is blocked by 25%, 50%, and 75%, respectively, a)-c) reveal the decryption of the Occlusion image

this encryption approach features lower computational complexity and a larger key space, effectively enhancing the system's resistance to potential attacks and improving image encryption performance. Future research can continue to develop the encryption algorithm and experimental simulations, such as using quantum neural networks to encrypt extracted ROIs and expanding the experimental dataset with more diverse images. To sum up, the selective encryption method based on quantum logarithm and morphology proposed in this study has verified the technical feasibility while ensuring efficiency and security through quantum circuit design and encryption strategy innovation. Its low complexity and large key space advantages provide a solution with both theoretical value and application potential for the field of image encryption, and the subsequent research directions also clarify the path for the continuous optimization and scene expansion of this technology.

Acknowledgments. The work is supported by the Shandong Provincial Natural Science Foundation (CN) (ZR2022LLZ003) and the National Natural Science Foundation of China (12471422).

Disclosure of Interests. The authors declare that they have no known competing financial interests or personal relationships that could have appeared to influence the work reported in this paper.

References

1. Tiken, C., Samli, R.: A comprehensive review about image encryption methods. Harran Üniversitesi Mühendislik Dergisi **7**(1), 27–49 (2022)
2. Ahmed, S.T., Hammood, D.A., Chisab, R.F., et al.: Medical image encryption: a comprehensive review. Computers **12**(8), 160 (2023)
3. Shor, P.W.: Algorithms for quantum computation: discrete logarithms and factoring. in: The 35th Annual IEEE Symposium on Foundations of Computer Science, Los Alamitos, CA, USA, pp. 124–134 (1994)
4. Grover, L.K.: A fast quantum mechanical algorithm for database search. In: The 28th Annual ACM Symposium on Theory of Computing, Philadelphia, pp. 212–219. ACM, USA (1996)
5. Nielsen, M.A., Chuang, I.L.: Quantum Computation and Quantum Information, pp. 157–223. Cambridge University Press, Cambridge (2005)
6. Wang, Z., Xu, M., Zhang, Y.: Review of quantum image processing. Arch. Comput. Methods in Eng. **29**(2), 737–761 (2022)
7. Kolivand, H., Hamood, S.F., Asadianfam, S., et al.: Image encryption techniques: a comprehensive review. Multimed. Tools Appl. 1–36 (2024)
8. Puech, W., Bors, A.G., Rodrigues, J.M.: Protection of color images by selective encryption. In: Advanced Color Image Processing and Analysis, pp. 397–421 (2013)
9. Norcen, R., Podesser, M., Pommer, A., et al.: Confidential storage and transmission of medical image data. Comput. Biol. Med. **33**(3), 277–292 (2003)
10. Nazneen, M.G., Banu, S., Tabassum, Z., et al.: Selective bitplane encryption for secure transmission of image data in mobile environment. Int. J. Sci. Technol. Res. **2**(6), 92–96 (2013)
11. Abd El-Latif, A.A., Abd-El-Atty, B., Talha, M.: Robust encryption of quantum medical images. IEEE Access **6**, 1073–1081 (2017)
12. Janani, T., Brindha, M.: A secure medical image transmission scheme aided by quantum representation. J. Inf. Secur. Appl. **59**, 102832 (2021)
13. Wei, Y., Li, H.S., Liu, K., et al.: Quantum video encryption based on bitplanes and improved Arnold scrambling. Quantum Inf. Process. **23**, 67 (2024)
14. Grigoryan, A.M., Agaian, S.S.: New look on quantum representation of images: Fourier transform representation. Quantum Inf. Process. **19**(5), 1–26 (2020). https://doi.org/10.1007/s11128-020-02643-3
15. Su, J., Guo, X., Liu, C., et al.: An improved novel quantum image representation and its experimental test on IBM quantum experience. Sci. Rep. **11**(1), 13879 (2021)
16. Nguyen, T., Paik, I., Sagawa, H., Thang, T.C.: Quantum machine learning with quantum image representations. In: 2022 IEEE International Conference on Quantum Computing and Engineering (QCE), Broomfield, CO, USA, pp. 851–854 (2022)
17. Ranjan, A., Arya, A.K.S., Ravinder, M.: Quantum techniques for image processing. In: 2020 2nd International Conference on Advances in Computing, Communication Control and Networking (ICACCCN), Greater Noida, India, pp. 1035–1039 (2020)
18. Beach, G., Lomont, C., Cohen, C.: Quantum image processing (quip). In: 2003 Proceedings of the 32nd Applied Imagery Pattern Recognition Workshop, pp. 39–44. IEEE (2003)
19. Le, P.Q., Dong, F., Hirota, K.: A flexible representation of quantum images for polynomial preparation, image compression, and processing operations. Quantum Inf. Process. **10**, 63–84 (2011)
20. Zhang, Y., Lu, K., Gao, Y., et al.: NEQR: a novel enhanced quantum representation of digital images. Quantum Inf. Process. **12**, 2833–2860 (2013)
21. Sun, B., Le, P.Q., Iliyasu, A.M., et al.: A multi-channel representation for images on quantum computers using the RGBα color space. In: IEEE International Symposium on Intelligent Signal Processing, Floriana, Malta, pp. 1–6 (2011)

22. Sun, B., Iliyasu, A., Yan, F., et al.: An RGB multi-channel representation for images on quantum computers. J. Adv. Comput. Intell. Intell. Inform. **17**(3) (2013)
23. Li, H.S., Zhu, Q., Zhou, R.G., et al.: Multi-dimensional color image storage and retrieval for a normal arbitrary quantum superposition state. Quantum Inf. Process. **13**, 991–1011 (2014)
24. Caraiman, S., Manta, V.: Image processing using quantum computing. In: IEEE International Conference on System Theory, Control and Computing (ICSTCC), Sinaia, Romania, pp. 1–6 (2012)
25. Kamiyama, M., Taguchi, A.: HSI color space with same gamut of RGB color space. IEICE Trans. Fundam. Electron. Commun. Comput. Sci. **100**(1), 341–344 (2017)
26. Kong, F., Peng, Y.: Color image watermarking algorithm based on HSI color space. In: 2010 2nd International Conference on Industrial and Information Systems, Dalian, China, pp. 464–467 (2010)
27. Saravanan, G., Yamuna, G., Nandhini, S.: Real time implementation of RGB to HSV/HSI/HSL and its reverse color space models. In: 2016 International Conference on Communication and Signal Processing (ICCSP), Melmaruvathur, India, pp. 0462–0466 (2016)
28. Taguchi, A., Hoshi, Y.: Color image enhancement in HSI color space without gamut problem. IEICE Trans. Fundam. Electron. Commun. Comput. Sci. **98**(2), 792–795 (2015)
29. Chen, G.L., Song, X.H., Venegas-Andraca, S.E., et al.: QIRHSI: novel quantum image representation based on HSI color space model. Quantum Inf. Process. **21**(1), 5 (2022)
30. Song, X., Chen, G., Abd El-Latif, A.A.: Quantum color image encryption scheme based on geometric transformation and intensity channel diffusion. Mathematics **10**(17), 3038 (2022)
31. Song, X., Chen, G., Abd El-Latif, A.A., et al.: Quantum geometric transformation based on QIRHSI quantum color images. IEEE Access **11**, 21883–21899 (2023)
32. Li, M., Song, X., El-Latif, A.A.A.: EQIRHSI: enhanced quantum image representation using entanglement state encoding in the HSI color model. Quantum Inf. Process. **22**(9), 334 (2023)
33. Xiao, D., Fu, Q., Xiang, T., et al.: Chaotic image encryption of regions of interest. Int. J. Bifurc. Chaos **26**(11), 1650193 (2016)
34. Ntalianis, K.S.: An automatic human video objects encryption scheme built on stream and block ciphers and based on chaos. In: Proceedings of the 9th World Scientific and Engineering Academy and Society (WSEAS) International Conference on Systems, Wisconsin, United States, pp. 1–7 (2005)
35. Liu, Y., Zhao, F., Xu, Y., et al.: A novel face template protection algorithm based on the fusion of chaos theory and RSA encryption. Int. J. Secur. Appl. **10**(6), 315–330 (2016)
36. Cheng, W., An, G.: Face template protection using chaotic encryption. In: The 5th IET International Conference on Wireless, Mobile and Multimedia Networks, Beijing, China, pp. 4–12 (2013)
37. Chen, R., Li, X., Teng, L., et al.: Selective region medical image encryption algorithm based on cascade chaos and two-dimensional joseph traversal. Phys. Scr. **98**(3), 035227 (2023)
38. Prakash, S.A., Kumar, A.G., Anandavel, L., et al.: Selective encryption using segmentation mask with chaotic Henon map for multidimensional medical images. arXiv preprint arXiv: 2403.04781 (2024)
39. Zhang, L., Song, X., El-Latif, A.A.A., et al.: Reversibly selective encryption for medical images based on coupled chaotic maps and steganography. Complex Intell. Syst. **10**(2), 2187–2213 (2024)
40. Heidari, S., Naseri, M., Nagata, K.: Quantum selective encryption for medical images. Int. J. Theor. Phys. **58**, 3908–3926 (2019)
41. Hu, H.: Research on region of interest extraction for quantum images and quantum 3D models. Beijing University of Technology (2020)
42. Li, M., Song, X., Zhao Y., Abd El-Latif, A.A.: Space-frequency based multichannel dual encryption for quantum color images using chaotic system and quantum walks. Quantum Inf. Process. (under review)

43. Alvarez, G., Li, S.: Some basic cryptographic requirements for chaos-based cryptosystems. Int. J. Bifurc. Chaos **16**(08), 2129–2151 (2006)
44. Wang, J., Song, X., Wang, H., et al.: Applicable image security based on new hyperchaotic system. Symmetry **13**(12), 2290 (2021)

Author Index

X. Zhou et al. (Eds.): GPC 2024, LNCS 15225, p. 229, 2026.
https://doi.org/10.1007/978-981-95-1346-8

MIX
Papier aus verantwortungsvollen Quellen
Paper from responsible sources
FSC® C105338

If you have any concerns about our products,
you can contact us on
ProductSafety@springernature.com

In case Publisher is established outside the EU,
the EU authorized representative is:
Springer Nature Customer Service Center GmbH
Europaplatz 3, 69115 Heidelberg, Germany

Printed by Libri Plureos GmbH
in Hamburg, Germany